Applied
Finite Element
Modeling

MECHANICAL ENGINEERING

A Series of Textbooks and Reference Books

Editor: L.L. FAULKNER   Columbus Division, Battelle Memorial Institute, and Department of Mechanical Engineering, The Ohio State University, Columbus, Ohio

Associate Editor: S.B. MENKES   Department of Mechanical Engineering, The City College of the City University of New York, New York

1. Spring Designer's Handbook, *by Harold Carlson*
2. Computer-Aided Graphics and Design, *by Daniel L. Ryan*
3. Lubrication Fundamentals, *by J. George Wills*
4. Solar Engineering for Domestic Buildings, *by William A. Himmelman*
5. Applied Engineering Mechanics: Statics and Dynamics, *by G. Boothroyd and C. Poli*
6. Centrifugal Pump Clinic, *by Igor J. Karassik*
7. Computer-Aided Kinetics for Machine Design, *by Daniel L. Ryan*
8. Plastics Products Design Handbook, Part A: Materials and Components; Part B: Processes and Design for Processes, *edited by Edward Miller*
9. Turbomachinery: Basic Theory and Applications, *by Earl Logan, Jr.*
10. Vibrations of Shells and Plates, *by Werner Soedel*
11. Flat and Corrugated Diaphragm Design Handbook, *by Mario Di Giovanni*
12. Practical Stress Analysis in Engineering Design, *by Alexander Blake*
13. An Introduction to the Design and Behavior of Bolted Joints, *by John H. Bickford*
14. Optimal Engineering Design: Principles and Applications, *by James N. Siddall*
15. Spring Manufacturing Handbook, *by Harold Carlson*
16. Industrial Noise Control: Fundamentals and Applications, *edited by Lewis H. Bell*
17. Gears and Their Vibration: A Basic Approach to Understanding Gear Noise, *by J. Derek Smith*

Additional Volumes in Preparation

Mechanical Engineering Software

Spring Design with an IBM PC, *by Al Dietrich*

Mechanical Design Failure Analysis: With Failure Analysis System Software for the IBM PC, *by David G. Ullman*

Applied
Finite Element
Modeling
Practical Problem Solving for Engineers

JEFFREY M. STEELE

Eastman Kodak Company
Rochester, New York

Marcel Dekker, Inc.   New York and Basel

Library of Congress Cataloging-in-Publication Data

Steele, Jeffrey M., [date]
  Applied finite element modeling.

  (Mechanical Engineering ; 66)
  Includes index.
  1. Finite element method. 2. Engineering--Mathematical
models. I. Title. II. Series: Mechanical engineering
(Marcel Dekker, Inc.) ; 66.
TA347.F5S74   1989   620'.001'515353   88-36332
ISBN 0-8247-8048-5 (alk. paper)

This book is printed on acid-free paper.

MARCEL DEKKER, INC.
270 Madison Avenue, New York, New York 10016

Current printing (last digit):
10 9 8 7 6 5 4 3 2 1

PRINTED IN THE UNITED STATES OF AMERICA

PREFACE

This book is intended to fill a gap in the information available to users of the finite-element method. The vast majority of finite element analyses is performed with commercial finite-element programs that provide documentation on coding input data and the mathematical basis for that particular program. In addition, there are a number of excellent textbooks on the theory and mathematical fundamentals of the finite element method. This book is intended to bridge the gap between the theoretical texts and the program user's manuals.

Practicing engineers, when faced with a real engineering problem to solve, need to answer questions such as "Is this problem applicable to finite element analysis?" "What type of model should be used?" "What type of elements and how many?" and "How accurate will the results be?" Extracting specific information from the theoretical texts to answer questions like these is difficult, especially for the novice.

As little as 10 years ago, finite element analysis was performed by specialists, trained at a graduate level, using mainframe computers. The rapid decrease in computing cost and the availability of engineering workstations and personal computers have brought along a new group of users who are not full-time analysts but use finite element analysis as another tool in their design and analysis work. These users may or may not have the benefit of formal training in the finite element method. It is for these users that this book is primarily intended. This book is not intended as a substitute for formal training, and a full college-level course is recommended for anyone seriously using the finite element method. This book provides guidance to the finite element user in building a cost-effective model and in thinking through the entire modeling process before sitting down at a computer terminal.

Although the price of mainframe hardware is decreasing, there is still a need for reasonable, cost-effective models. Good results cannot be achieved by simple overkill. In addition, most hardware still has limitations, especially workstations and personal computers.

Chapter 2 provides an overview of the mathematical fundamentals of the finite element method and gives references to a few texts recommended for more complete information. Topics for inclusion in this chapter were selected to be consistent with the material covered in the following chapters. Although topics such as substructuring and dynamics may be considered advanced by some standards, they are in common use and are therefore included.

Chapters 3, 4, 5, and 6 cover the basics of modeling. Chapter 3 on required input data is for the first-time user. Chapters 4 and 5 walk the reader through model development as a three-step process: (1) problem definition, (2) free-body diagram, and (3) finite element model. A set of simple components is included in Chapter 4 and carried through the subsequent chapters as example problems.

Chapters 7 and 8 discuss performance of the various element types, including the effects of element distortion. Most of the problems in these chapters were provided by a paper proposing a set of finite element benchmark problems by R. H. MacNeal and R. L. Harder of the MacNeal Schwendler Corporation. Data in these chapters were generated with MSC/NASTRAN and ANSYS. The elements are described as generically as possible without referencing the program or version.

Chapters 9 through 12 deal with more advanced topics such as refined mesh modeling for stress concentration effects, substructuring, dynamics, and thermal analysis.

Chapter 13 provides some ideas for assessing the accuracy of finite element analyses. Rather than be a set of rigid guidelines, this chapter is intended as food for thought and will, I hope, stir some interest in a topic that has not received as much attention as deserved.

I would like to express my appreciation to Dr. Neville F. Rieger, my colleague and former employer, for suggesting the topic for the book over 12 years ago. Also, I would like to thank the MacNeal Schwendler Corporation for the use of their set of benchmark problems and data that forms the basis for much of Chapters 7 and 8; the people of Swanson Analysis Systems, Inc., with whom I have enjoyed a good working relationship over the past eight years; and the editors of *Computer Aided Engineering* for publishing "A Manager's Guide to Finite Element Analysis." This series of articles served as a testing ground for many of the ideas that later found their way into this book. And finally, I express my gratitude to my wife Julie for her patience and support through all the hours of computer runs and typing.

Jeffrey M. Steele

CONTENTS

Applied
Finite Element
Modeling

1

INTRODUCTION TO THE FINITE ELEMENT METHOD

The finite element method of engineering analysis is relatively new and can trace its roots back to the mid-1950s. The paper that signaled the beginning of finite element analysis was written in 1956 by Turner et al. [1]. The finite element method requires the use of modern digital computing equipment with floating-point capability and considerable memory and disk capacity. For this reason, the development and popularity of the method has closely followed the evolution of computing hardware over the past 30 years.

The finite element method has been applied to a wide variety of engineering problems in which complex geometry has not allowed the use of a simple closed-form solution. The method's versatility and popularity stem from the use of simple shapes, known as elements, to be assembled together to model geometrically complex structures.

The first major finite element code for general use was NAS-TRAN developed for NASA by the MacNeal—Schwendler Corporation and Computer Sciences Corporation in the mid-1960s. As with the early work of Turner et al., it had its roots in the aerospace industry. The method was applied to other areas of structural analysis in mechanical and civil engineering. The ANSYS code had its roots in the nuclear industry and was introduced in 1970 by Swanson Analysis Systems Inc. Zienkiewicz published his text [2] in 1971 with emphasis on civil as well as mechanical engineering applications.

A significant step in finite element development occurred in 1968 with the paper by Irons and Zienkiewicz that advanced the isoparametric element concept [3]. Another significant contribution was the introduction of the frontal solution algorithm by Irons in 1970 [4]. The frontal solution routine made the finite element method more adaptable to a wider variety of hardware.

The finite element method has been expanded to nonstructural field problems such as heat transfer, fluid flow, and magnetics. In 1965, Zienkiewicz and Cheung's findings appeared [5] and in 1966, Wilson and Nickell's [6].

In the late 1960s and into the 1970s, the application of the finite element method required the use of a large mainframe computer. Most applications were run on IBM and Control Data machines that had memory and precision sufficient to handle the large sets of matrix equations generated during a calculation. The majority of the analyses were performed by specialists with training in applied mechanics, finite element methods, use of computer software, and use of the finite element code. Input data had to be carefully coded because there was a significant cost per second on the computer and errors in data input, or data entry could result in significant wasted costs.

In the late 1970s, the introduction of super-minicomputers such as the Digital VAX, the Prime, and the Data General Eclipse made it possible to bring the necessary hardware into the engineering department. With these machines, it was possible for design and analysis engineers to use finite element analysis as part of their work without necessarily relying on the support of a finite element specialist. The development of engineering workstations in the early and mid-1980s such as the Apollo, the Sun, and the Microvax helped to further promote the use of finite element applications. With the cost per second of computing use decreasing at a rapid rate, the cost of finite-element analysis decreased to a point where engineers were not restricted by computer costs. The development of the Intel 80286/80287-based PC/AT in the mid-1980s and its clones made possible the porting of some finite element codes onto a desktop machine. Although the PC machines have fixed memory and do not have their own virtual memory capability, they are still capable of running modest finite element problems. It is only a matter of time until the engineer will have a desktop machine, with a full 32-bit capability and virtual memory at his disposal. This entire hardware evolution has made it possible for the average engineer to have the capability to use finite element analysis as another design and analysis tool.

The objective of this book is to bridge the gap between the excellent texts that have been written on the finite element method [2, 7–10] and the user's data input instructions for the various finite element codes. The intention of this book is to give the novice or part-time finite element user information on the modeling process in general. Information is presented in a generic sense without tying it directly to one or another specific finite-element code. A general overview of the basis for the finite element method is given in Chapter 2; however, it is not intended as a substitute for the more comprehensive texts given above.

Subsequent chapters give guidelines and recommendations for going from an engineering problem to a finite element calculation including defining the problem, developing the most cost-effective model, and interpreting the results. Although there are many finite element codes available, each with a different data input scheme, there is a common set of basic input data required for each. The objective of an engineer using finite element analysis is to work from the engineering problem to develop a model that will give him suitable accuracy at a reasonable price in terms of analysis and computer time.

REFERENCES

1. Turner, M. J., R. W. Clough, H. C. Martin, and L. J. Topp, "Stiffness and Deflection Analysis of Complex Structures," *Journal of Aeronautical Science*, Vol. 23, 1956, pp. 805–824.
2. Zienkiewicz, O. C., *The Finite Element Method in Engineering Science*, 2nd ed., McGraw Hill, London, 1971.
3. Irons, B. M. and O. C. Zienkiewicz, "The Isoparametric Finite Element System — A New Concept in Finite Element Analysis," Proceedings of the Conference on Recent Advances in Stress Analysis, Royal Aeronautics Society, 1968.
4. Irons, B. M., "A Frontal Solution Program for Finite Element Analysis," *International Journal for Numerical Methods in Engineering*, Vol. 2, No. 1, Jan. 1970, pp. 5–23.

5. Zienkiewicz, O.C. and Y. K. Cheung, "Finite Elements in the Solution of Field Problems," *The Engineer*, Sept. 1965, pp. 507–10.

6. Wilson, E. L. and R. E. Nickell, "Application of the Finite Element Method to Heat Conduction Analysis," *Nuclear Engineering and Design*, Vol. 4, 1966, pp. 276–286.

7. Cook, R. D., *Concepts and Applications of Finite Element Analysis*, Wiley, New York, 1974.

8. Gallagher, R. H., *Finite Element Analysis Fundamentals*, Prentice Hall, Englewood Cliffs, N.J., 1975.

9. Heubner, K. H., *The Finite Element Method for Engineers*, Wiley, New York, 1975.

10. Bathe, K. J. and E. L. Wilson, *Numerical Methods in Finite Element Analysis*, Prentice-Hall, Englewood Cliffs, N.J., 1976.

2

FUNDAMENTALS OF THE FINITE ELEMENT METHOD

The mathematics of the finite element method can be complex. This chapter gives a broad overview of the fundamentals of the method, but does not attempt to give a rigorous treatment of the mathematics. There are many excellent texts on the fundamentals of the finite element method and the reader is referred to them as sources of basic information. They include [1, 3, 4, 7−9].

2.1 TERMINOLOGY

The finite element method operates on the assumption that any continuous function over a global domain can be approximated by a series of functions operating over a finite number of small sub-domains. These series of functions are piecewise continuous and should approach the exact solution as the number of subdomains approaches infinity.

1. The global domain is divided into subdomains called elements.
2. The points defining and connecting the elements are called *nodes* or *nodal points.*
3. The function that exists over the domain is explicitly solved for at the nodal points, i.e., nodal variables. It is assumed that the value of the function at any point internal to an element can be defined in terms of that element's nodal variables. The nodal variables are referred to as *degrees of freedom.* This term applies specifically to stress analysis in which the nodal variables are the deflections of the structure at the nodal points; however, the term is often used generically to refer to all nodal variables.

Although the elements are specified as joined at their common nodes, they are assumed to be continuously coupled along their common boundary and any function is assumed to be continuous at the boundaries, although continuity of slope is not necessarily maintained. The complete collection of elements represents an approximation of the domain's geometry as a continuum. The nodal points are only reference points for evaluation of the function and do not represent physical points of connection within the domain.

2.2 ELEMENT DISCRETIZATION

The concept of element discretization can be illustrated by a finite element model of a tapered beam in tension as shown in Figure 2.1. The domain is the two-dimensional plane of the beam. The function to be evaluated is the displacement field in the axial direction. There is a force in that direction, and one degree of freedom per node. The beam is divided into three elements and four nodes. A force is applied to node 4, and node 1 is constrained against having any displacement.

The stiffness functions for each of the three elements can be formed by the relationship

$$k = EA/L \qquad\qquad (2.1)$$

where

k = stiffness lbf/in.^{2}
E = Young's modulus
A = the average cross-sectional area of the element
L = the length of the element

The three element stiffnesses may be calculated, assuming unit thickness and $E = 30,000,000$ psi, as

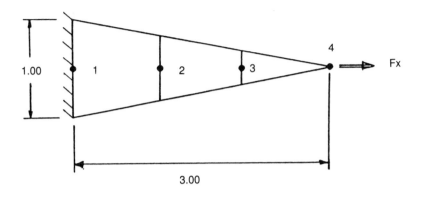

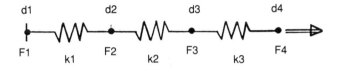

Resulting Equivalent Spring Model

Figure 2.1 Finite-element model of a tapered beam: Beam in tension divided into three one-dimensional elements.

Element 1 A_{ave} = 0.167 in.^{2} k_1 = 5,000,000 lbf/in.

Element 2 A_{ave} = 0.500 in.^{2} k_2 = 15,000,000 lbf/in.

Element 3 A_{ave} = 0.833 in.^{2} k_3 = 25,000,000 lbf/in.

The nodal variables to be solved for, i.e., the degrees of freedom, are the axial displacements of each of the nodes d_i. Assembling a set of equations to represent the beam gives

Equation 1 $k_1(d_1 - d_2) = F_1 - F_2$ (2.2)

Equation 2 $k_2(d_2 - d_3) = F_2 - F_3$

Equation 3 $k_3(d_3 - d_4) = F_3 - F_4$

where

d_i = axial displacement of node i

F_i = axial force at node i

Writing these equations in matrix form for element 1 gives

$$\begin{bmatrix} k_1 & -k_1 \\ -k_1 & k_1 \end{bmatrix} \begin{Bmatrix} d_1 \\ d_2 \end{Bmatrix} = \begin{Bmatrix} F_1 \\ F_2 \end{Bmatrix} \tag{2.3}$$

The matrix set of stiffness equations for the other elements is of the same form.

The assembled set of matrix equations describing the entire beam is

$$\begin{bmatrix} k_1 & -k_1 & & \\ -k_1 & k_2 + k_3 & -k_2 & \\ & -k_2 & k_2 + k_3 & -k_3 \\ & & -k_3 & k_3 \end{bmatrix} \begin{Bmatrix} d_1 \\ d_2 \\ d_3 \\ d_4 \end{Bmatrix} = \begin{Bmatrix} F_1 \\ F_2 \\ F_3 \\ F_4 \end{Bmatrix} \tag{2.4}$$

This set of matrix equations represents four equations in four unknowns, d_i, which can be solved to characterize the response of the beam to an axial applied force.

Although this example illustrates the principle of assembling element stiffness equations into a global set of equations, actual formulation of the individual element stiffness matrices for two-dimensional and three-dimensional solid elements is significantly more complex. Some advanced three-dimensional solid elements have 20 nodes and 60 degrees of freedom per element represented by a 60 × 60 set of equations for each element.

2.3 ISOPARAMETRIC CONCEPT

In order to develop stiffness equations for two-dimensional and three-dimensional solid elements, there need to be interpolation functions that will give the value of a variable at an interior point of the element as a function of the nodal values. These interpolation functions are typically referred to as shape functions. For an element in an x, y coordinate system, the displacement at some element interior point may be specified as

$$d(x, y) = \sum_{i=1}^{4} N_i(x, y) d_i \tag{2.5}$$

where

$d(x, y)$ = displacement at global coordinates x, y
d_i = nodal displacement values
$N_i(x, y)$ = shape functions for each node and a function of x and y

In most practical application of the finite element method, elements will not have perfectly square, cubic, or triangular shapes. In order to develop elements that can give accurate results for an arbitrary shape, there needs to be a method of specifying the governing equations in a local coordinate system and mapping these equations into the actual element's distorted shape. The elements in the local coordinate system are perfectly square or cubic and are defined by the local coordinate system as shown in Figure 2.2. These local elements can then be mapped into a global coordinate system by use of a suitable set of mapping functions as shown in Figure 2.3.

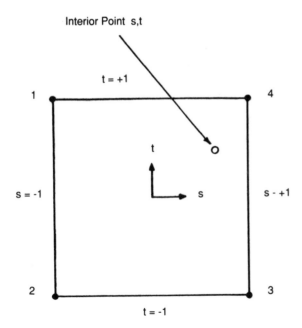

Interior Points May Be Described with Local Coordinates: s,t

Figure 2.2 Local element coordinate system.

ELEMENT IN GLOBAL COORDINATE SYSTEM

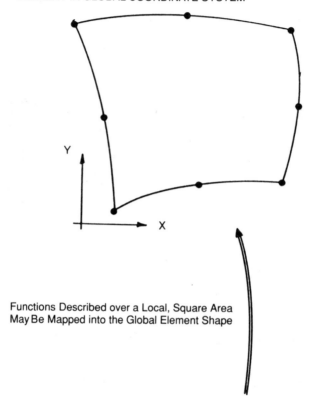

Functions Described over a Local, Square Area
May Be Mapped into the Global Element Shape

ELEMENT IN LOCAL COORDINATE SYSTEM

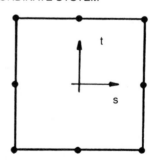

Figure 2.3 Local to global element mapping.

In this manner, both two- and three-dimensional elements can be mapped into any reasonable quadrilateral or prism shape. More important, when the elements have midside nodes in addition to their corner nodes, the elements can match a curved boundary in the global system. The global coordinates for any point in the element can be specified in terms of that point's local coordinates and the global coordinates of the nodes

$$x(s, t) = \sum_{i=1}^{n} N_i'(s, t) \, x_i \qquad (2.6)$$

where

$x(s, t)$ = global coordinate of point s, t
n = the number of nodes
$N_i'(s, t)$ = mapping function for node i
x_i = x coordinate for node i

In the same manner, the y coordinate is

$$y(s, t) = \sum_{i=1}^{n} N_i'(s, t) \, y_i \qquad (2.7)$$

The definition of an *isoparametric* element is one that uses the same functions to map the coordinates as it uses to interpolate nodal variables, i.e.,

$$N'(s, t) = N(s, t) \qquad (2.8)$$

The isoparametric concept is thoroughly discussed by Zienkiewicz in [1].

2.4 ISOPARAMETRIC SHAPE FUNCTIONS

For a two-dimensional quadrilateral with corner nodes as shown in Figure 2.4, the isoparametric shape functions can be given by

$N_1 = 1/4(1-s)(1-t)$

$N_2 = 1/4(1+s)(1-t)$

$$(2.9)$$

$N_3 = 1/4(1+s)(1+t)$

$N_4 = 1/4(1-s)(1+t)$

A Function May Be Evaluated at an Interior Point of an Element
by Summing the Value of the Function at the Nodal Points,
Multiplied by the Appropriate Weighting Factors

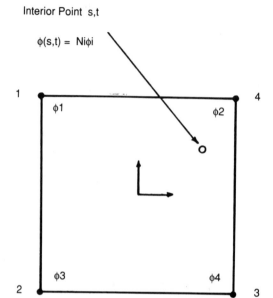

Figure 2.4 Quadrilateral element.

Several items should be noted regarding these shape functions.

They are linear polynomial functions. Both the coordinates
and nodal variables are linearly interpolated within the element. This can be better illustrated by considering interpolation in the local s direction along the "top" surface of the
element (Figure 2.5), i.e., t = 1 between nodes 3 and 4 where
the shape functions become

$$N_3 = 1/2(1+s)$$

$$N_4 = 1/2(1-s)$$
(2.10)

The shape function for a given node i has a value of 1.0 at
node i and a value of 0.0 at each of the other nodes.

At any given local point (s, t), the sum of all four shape functions must equal 1.0.

$$\phi = N3\phi3 + N4\phi4$$

$$= 0.5(1+s) + 0.5(1-s)$$

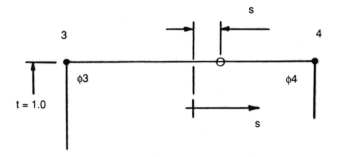

For One Dimension the Value of the Function at the Interior
Point Is the Weighted Average of the Two Corner Node Values
The Weighting Functions Are Linear and Based on the Distance
from the Interior Point to the Node

Figure 2.5 One-dimensional linear interpolation.

2.5 STRESS, STRAIN, AND STIFFNESS FORMULATIONS

Given the two-dimensional element shown in Figure 2.4, the strain
at any point in the element can be given by the strain vector

$$\{\varepsilon\} = \begin{Bmatrix} \varepsilon_x \\ \varepsilon_y \\ \varepsilon_{xy} \end{Bmatrix} \qquad (2.11)$$

where

ε_x = strain in the x direction = du/dx
ε_y = strain in the y direction = dv/dy
ε_{xy} = shear strain in the x-y plane = $du/dy + dv/dx$

The shape functions are used to obtain these derivatives with
respect to the nodal displacements. For example,

$$\varepsilon_x = du/dx = d/dx \left[\sum_{i=1}^{n} N_i u_i \right] \qquad (2.12)$$

The y direction strain and shear strain can be obtained in a similar manner.

The complete calculation of strain can be given in matrix form as

$$\{\varepsilon\} = \begin{bmatrix} dN_1/dx...dN_n/dx & 00...000 \\ 000...00 & dN_1/dy...dN_n/dy \\ dN_1/dy...dN_n/dy & dN_1/dx...dN_n/dx \end{bmatrix} \begin{Bmatrix} u_1 \\ ... \\ u_n \\ v_1 \\ ... \\ v_n \end{Bmatrix}$$

The matrix of shape function global derivatives is known as the [B] matrix and is also referred to as the strain-displacement matrix. The above set of equations can be given in matrix form as

$$\{\varepsilon\} = [B]\{d\} \tag{2.13}$$

Stress and strain can be related by the use of an elasticity matrix [D].

$$\{\sigma\} = [D]\{\varepsilon\} \tag{2.14}$$

For two-dimensional plane stress and isotropic materials, the elasticity matrix is

$$[D] = \frac{E}{1-\mu^2} \begin{bmatrix} 1 & \mu & 0 \\ \mu & 1 & 0 \\ 0 & 0 & (1-\mu)/2 \end{bmatrix} \tag{2.15}$$

For two-dimensional plane strain and isotropic materials, the elasticity matrix is

$$[D] = \frac{E(1-\mu)}{(1+\mu)(1-2\mu)} \begin{bmatrix} 1 & \mu/(1-\mu) & 0 \\ \mu/(1-\mu) & 1 & 0 \\ 0 & 0 & (1-2\mu)/2(1-\mu) \end{bmatrix} \tag{2.16}$$

In order to convert nodal displacements into stresses, the B and D matrices are combined.

$$\{\sigma\} = [D][B]\{d\} \tag{2.17}$$

It should be noted that the above relationships hold at one specific point within the element as defined by the local coordinates used in the shape functions N_i whose derivatives are in [B].

The element stiffness matrix for a two-dimensional element is generated by integrating the product of the transpose of the B matrix, the D matrix, and the B matrix over the element area (or volume in the case of three-dimensional solid elements), i.e.,

$$[k] = \int [B]^T [D][B] \, dx \, dy \qquad (2.18)$$

where [k] is the element stiffness matrix and the relationship between stiffness, applied forces, and nodal displacements is given by

$$[k]\{d\} = \{F\} \qquad (2.19)$$

2.6 ELEMENT EQUATION ASSEMBLY AND SOLUTION

There are two methods commonly used to assemble and solve the element matrix sets of equations: the banded equation solver and the wavefront equation solver.

In the banded solver method, the entire global set of equations from all elements is first assembled and then reduced. The method of assembling the individual element stiffness matrices into a global stiffness matrix follows the same form as shown in the simple example in Section 2.2 and in Equation 2.4.

The element force and displacement vectors, and stiffness coefficients are set up according to the order of the global degrees of freedom. For example, if a particular element is bounded by nodes 1, 2, 10, 9 (degrees of freedom 1, 2, 3, 4, 19, 29, 17, 18) as shown in Figure 2.6, then the element stiffness coefficient in row 1 and column 7 represents the relationship between a force at local degree of freedom 1 (1st node, x direction) and a displacement at local degree of freedom 7 (4th node, x direction).

$$\begin{bmatrix} k_{1,1} & k_{1,2} & k_{1,3} & k_{1,4} & k_{1,5} & k_{1,6} & k_{1,7} & k_{1,8} \\ k_{2,1} & k_{2,2} & k_{2,3} & k_{2,4} & k_{2,5} & k_{2,6} & k_{2,7} & k_{2,8} \\ & & & \cdots & & & & \\ k_{8,1} & k_{8,2} & k_{8,3} & k_{8,4} & k_{8,5} & k_{8,6} & k_{8,7} & k_{8,8} \end{bmatrix} \begin{Bmatrix} d_1 \\ d_2 \\ \cdots \\ d_8 \end{Bmatrix} = \begin{Bmatrix} F_1 \\ F_2 \\ \cdots \\ F_8 \end{Bmatrix} \qquad (2.20)$$

In the global system, this relates node 1, x direction and node 9, x direction (degree of freedom 17) so that the coefficient would be assembled into the global matrix at row 1, column 17.

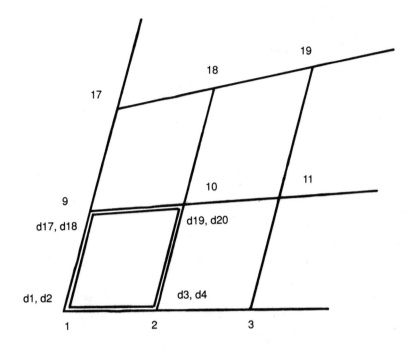

Figure 2.6 Typical two-dimensional element gridwork.

$$
\begin{bmatrix}
k_{1,1} & k_{1,2} & k_{1,3} & k_{1,4} & k_{1,19} & k_{1,20} & k_{1,17} & k_{1,18} \\
k_{2,1} & k_{2,2} & k_{2,3} & k_{2,4} & k_{2,19} & k_{2,20} & k_{2,17} & k_{2,18} \\
\cdots & & & & & & & \\
\cdots & & & & & & & \\
k_{18,1} & k_{18,2} & k_{18,3} & k_{18,4} & k_{18,19} & k_{18,20} & k_{18,17} & k_{18,18}
\end{bmatrix}
\begin{Bmatrix}
d_1 \\ d_2 \\ \cdots \\ \cdots \\ d_{18}
\end{Bmatrix}
=
\begin{Bmatrix}
F_1 \\ F_2 \\ \cdots \\ \cdots \\ F_{18}
\end{Bmatrix}
$$

(2.21)

Two properties of the stiffness matrix are significant with respect to the solution algorithm.

1. The matrix is symmetric, i.e., $k_{1,17} = k_{17,1}$.
2. The matrix is positive-definite, i.e., there are no negative
    terms in the matrix and no zero terms on the matrix diag-
    onal.

Rather than finding an inverse of the stiffness matrix, trian-
gularization is used to reduce the matrix to an upper triangular

form, meaning that all terms below the diagonal are 0. The original set of matrix stiffness equations may look like this

$$
\begin{bmatrix}
k_{11} & k_{12} & k_{13} & k_{14} & k_{15} \\
k_{21} & k_{22} & k_{23} & k_{24} & k_{25} \\
k_{31} & k_{32} & k_{33} & k_{34} & k_{35} \\
k_{41} & k_{42} & k_{43} & k_{44} & k_{45} \\
k_{51} & k_{52} & k_{53} & k_{54} & k_{55}
\end{bmatrix}
\begin{Bmatrix}
d_1 \\ d_2 \\ d_3 \\ d_4 \\ d_5
\end{Bmatrix}
=
\begin{Bmatrix}
F_1 \\ F_2 \\ F_3 \\ F_4 \\ F_5
\end{Bmatrix}
\tag{2.22}
$$

and is reduced to look like this

$$
\begin{bmatrix}
k'_{11} & k'_{12} & k'_{13} & k'_{14} & k'_{15} \\
0 & k'_{22} & k'_{23} & k'_{24} & k'_{25} \\
0 & 0 & k'_{33} & k'_{34} & k'_{35} \\
0 & 0 & 0 & k'_{44} & k'_{45} \\
0 & 0 & 0 & 0 & k'_{55}
\end{bmatrix}
\begin{Bmatrix}
d_1 \\ d_2 \\ d_3 \\ d_4 \\ d_5
\end{Bmatrix}
=
\begin{Bmatrix}
F'_1 \\ F'_2 \\ F'_3 \\ F'_4 \\ F'_5
\end{Bmatrix}
\tag{2.23}
$$

The displacements can be obtained from this set of equations by back-substitution. The last equation contains only one unknown d_5 that can be explicitly solved for, i.e.,

$$
d_5 = F'_5 / k'_{5,5} \tag{2.24}
$$

The second to the last equation has two unknowns. However, one of those is d_5. Once d_5 has been solved for, the second to the last equation can be solved for d_4. The process continues back through the set of equations until all unknowns have been solved for.

In a large set of matrix equations, the stiffness matrix is said to be banded, i.e., the nonzero terms are clustered about the diagonal and the corners filled with zero terms. The simple example in Section 2.2 shows this. The reason for the off-diagonal zero terms is that stiffness coefficients only exist for degrees of freedom sharing common elements. There are no stiffness coefficients relating degrees of freedom at extremes of the model. In the simple matrix (2.3), note that there is no stiffness coefficient relating degrees of freedom 1 and 4. These degrees of freedom are related in the global sense via degrees of freedom 2 and 3 and through element 2, but there is no direct coupling. For that reason, there are zeros at locations 1, 4 and 4, 1 in the matrix. The same

```
000000000000000KKKKKKKKKKKKKKKKKKKKKKK0000000000000000000     d     F
000000000000000KKKKKKKKKKKKKKKKKKKKKKK0000000000000000000     d     F
00000000000000000000KKKKKKKKKKKKKKKKKKKKKKK00000000000000     d     F
00000000000000000000KKKKKKKKKKKKKKKKKKKKKKK00000000000000     d     F
0000000000000000000000000KKKKKKKKKKKKKKKKKKKKKKK000000000     d     F
0000000000000000000000000KKKKKKKKKKKKKKKKKKKKKKK000000000     d     F
000000000000000000000000000KKKKKKKKKKKKKKKKKKKKKKKK0000000     d     F
000000000000000000000000000KKKKKKKKKKKKKKKKKKKKKKKK00000     d     F
0000000000000000000000000000000KKKKKKKKKKKKKKKKKKKK00000000     d     F
0000000000000000000000000000000KKKKKKKKKKKKKKKKKKKK00000000     d     F
00000000000000000000000000000000000KKKKKKKKKKKKKKKKKKKKKK000     d     F
00000000000000000000000000000000000KKKKKKKKKKKKKKKKKKKKKK000     d     F
0000000000000000000000000000000000000KKKKKKKKKKKKKKKKKKKKKKK     d     F
0000000000000000000000000000000000000KKKKKKKKKKKKKKKKKKKKKKK     d     F
000000000000000000000000000000000000000000000KKKKKKKKKKKKKK     d     F
000000000000000000000000000000000000000000000000KKKKKKKKKK     d     F
```

Figure 2.7 Banded stiffness matrix.

principle applies to larger sets of equations in which the nonzero
terms may account for only 20—30% of the entire stiffness matrix
(Figure 2.7). The banded matrix equation reduction routines take
advantage of this and operate only within the nonzero band. Node
numbering becomes important to keep the bandwidth to a minimum.

The bandwidth for any given row (degree of freedom) is determined by the maximum and minimum degree of freedom numbers for
the elements to which the given degree of freedom connected. It
is therefore important in constructing a model for use with a
banded equation solver to minimize the difference in node numbers for each element (Figure 2.8).

The frontal method was first developed by Irons in 1970. This
method performs the assembly and solution phases simultaneously.
It has the advantage of not requiring as much computer memory
as the banded solver because the complete set of equations is
never actually assembled at any time. The frontal method loops
over the elements and checks each degree of freedom associated
with that element. Degrees of freedom that are current or that
will appear in a later element are retained in the wave and degrees of freedom that are making their last appearance are reduced out. The important parameter in setting up a model for
use with a frontal equation solver is element numbering. The
parameter to be minimized is the difference in element number
for elements sharing a common node (Figure 2.9). This is slightly
different from the case of the banded solver in which the parameter

Efficient Node Numbering

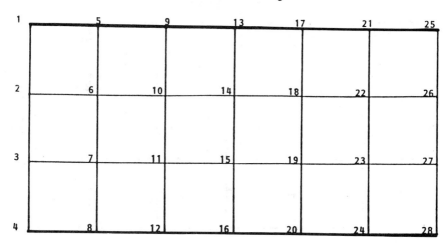

Numbering Should Minimize the Difference
in Node Numbers for Any Element

Less Efficient Node Numbering

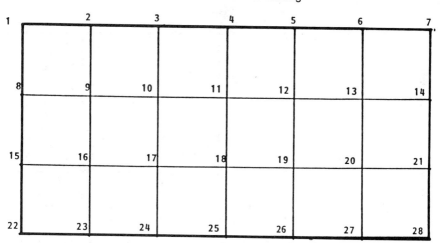

Figure 2.8 Banded equation solver numbering.

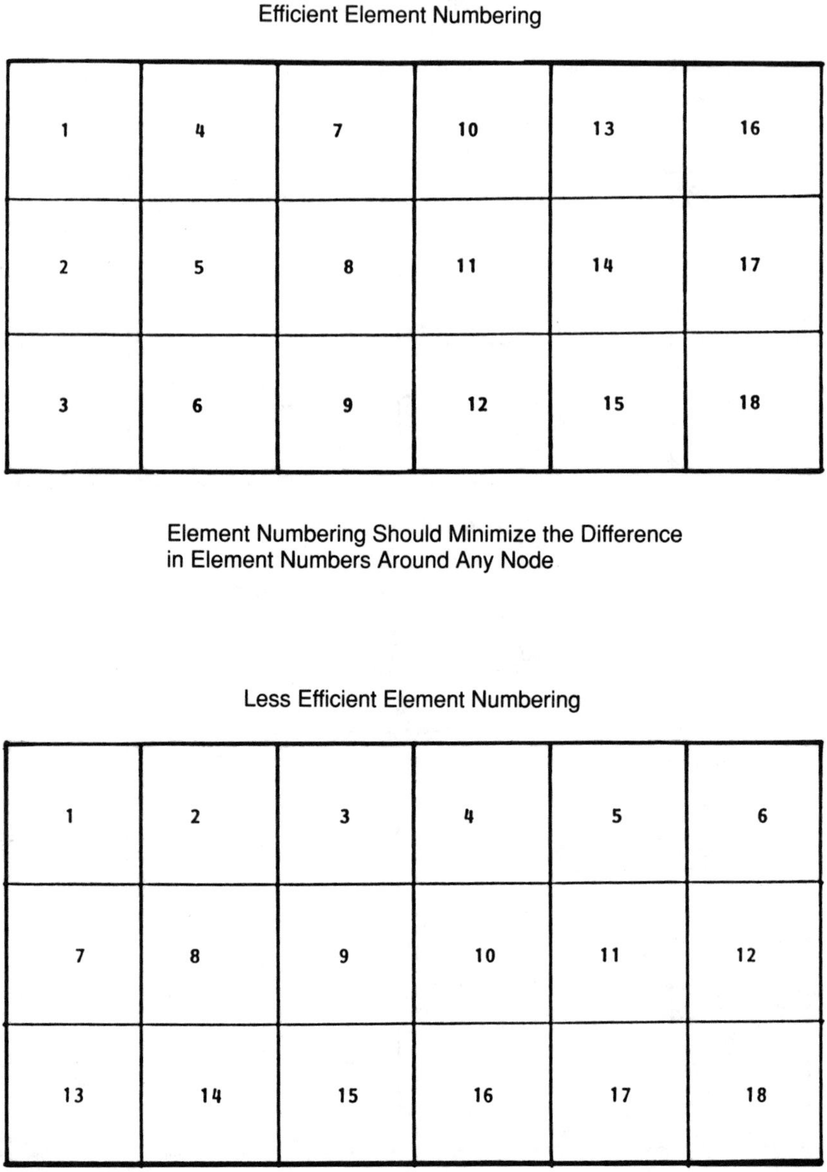

Efficient Element Numbering

1	4	7	10	13	16
2	5	8	11	14	17
3	6	9	12	15	18

Element Numbering Should Minimize the Difference
in Element Numbers Around Any Node

Less Efficient Element Numbering

1	2	3	4	5	6
7	8	9	10	11	12
13	14	15	16	17	18

Figure 2.9 Wavefront equation solver element numbering.

to be minimized is the difference in node number for nodes bounding an element.

2.7 ELEMENT TYPES

Element types can be broken down into a few basic groups, two-dimensional, three-dimensional solid, beam, and plate. Other specialty elements are spring, concentrated mass, gap, and damper elements.

2.7.1 Two-Dimensional Elements

Two-dimensional elements can be either plane stress, plane strain, or axisymmetric. Two-dimensional elements may be used when all forces and displacements act in plane. These elements have two degrees of freedom per node. Shapes include quadrilaterals and triangles. Elements may have nodes only at their vertices or they may have additional midside nodes. Axisymmetric elements are used to model solids of revolution such as pressure vessels.

Axisymmetric elements are classified as two-dimensional because they have only two degrees of freedom per node, displacements in the axial and radial directions. These elements, however, are capable of calculating out-of-plane strains and stresses in the tangential or hoop direction. Details of the two-dimensional element's stiffness matrix, and stress and strain relationships have been given in Sections 2.4 and 2.5.

2.7.2 Three-Dimensional Solid Elements

Three-dimensional solid elements are formulated as a direct extension of the two-dimensional elements described in Sections 2.4, 2.5, and 2.6. Three-dimensional solid elements have three degrees of freedom per node: translations in the x, y, and z directions. There are six strains and stresses calculated by these elements

$$
\{\varepsilon\} = \left\{ \begin{array}{c} \varepsilon_x \\ \varepsilon_y \\ \varepsilon_z \\ \varepsilon_{xy} \\ \varepsilon_{yz} \\ \varepsilon_{xz} \end{array} \right\}, \quad \{\sigma\} = \left\{ \begin{array}{c} \sigma_x \\ \sigma_y \\ \sigma_z \\ \sigma_{xy} \\ \sigma_{yz} \\ \sigma_{xz} \end{array} \right\} \tag{2.25}
$$

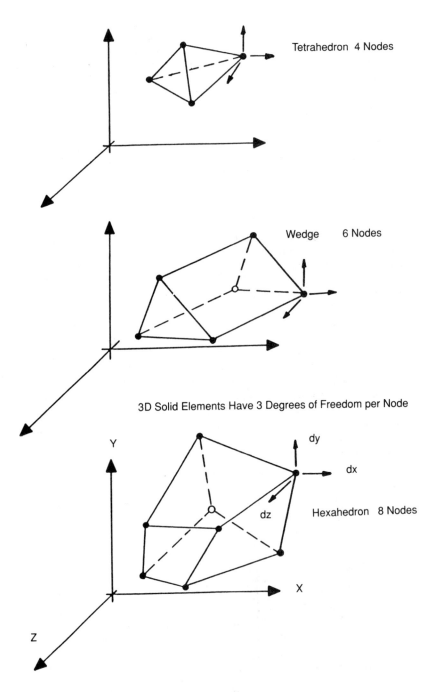

Tetrahedron 4 Nodes

Wedge 6 Nodes

3D Solid Elements Have 3 Degrees of Freedom per Node

dy

dx

dz

Hexahedron 8 Nodes

Y

X

Z

Figure 2.10 Typical three-dimensional solid element types.

Shapes include tetrahedra, wedge shapes, and rectangular prisms as shown in Figure 2.10. As in the two-dimensional cases, elements may have nodes only at their vertices or at their vertices and along their midsides.

2.7.3 Beam Elements

Beam elements have only one node at each end but have rotational degrees of freedom in order to transfer moments as well as forces (Figure 2.11). The six degrees of freedom per node for a beam

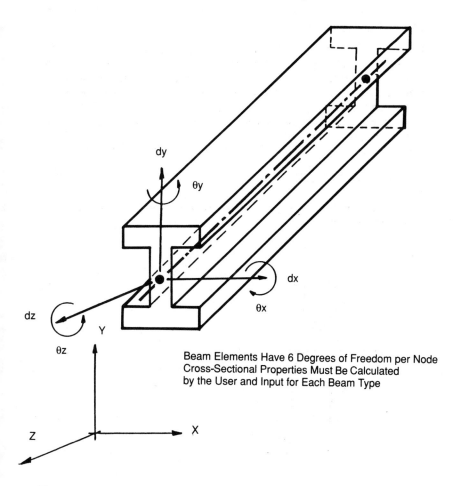

Beam Elements Have 6 Degrees of Freedom per Node
Cross-Sectional Properties Must Be Calculated
by the User and Input for Each Beam Type

Figure 2.11 Beam element.

element are the three displacements plus three rotations or slopes. Forces at the nodes consist of three forces and three moments. Beam elements assume constant or linearly varying cross-sectional properties. Properties such as cross-sectional area and area moments of inertia must be input for beam elements because the beam's geometry cannot be determined from the two nodes.

The stiffness matrix for a beam element is not formed by integration of stiffness properties over the volume, but the stiffness coefficients are calculated directly using closed-form procedures. An example of a beam element stiffness matrix is given in Table 2.1.

2.7.4 Plate Elements

Plate and shell elements also have six degrees of freedom per node and are a counterpart to the beam element (Figure 2.12). Most plate elements have only one node at their vertices so that the thickness of the plate must be specified either as a constant or linear variation.

The element stiffness matrix is formed by numerical integration over the element volume in a manner similar to the two-dimensional and three-dimensional elements discussed previously.

$$[k] = \int [B^T][D][B] \ d \ vol. \tag{2.26}$$

Displacements normal to the plate are defined as w_i so that the strain vector is

$$\{\varepsilon\} = \left\{ \begin{array}{c} - \dfrac{\partial^2 w}{\partial x^2} \\[2ex] - \dfrac{\partial^2 w}{\partial y^2} \\[2ex] 2 \dfrac{\partial^2 w}{\partial x \partial y^2} \end{array} \right\} \tag{2.27}$$

The corresponding displacement matrix consists of one set of displacements and two sets of rotations (slopes).

$$\{d\} = \left\{ \begin{array}{c} w_i \\[1ex] \theta_{xi} \\[1ex] \theta_{yi} \end{array} \right\} = - \left\{ \begin{array}{c} w_i \\[2ex] \dfrac{\partial w}{\partial y_i} \\[2ex] \dfrac{\partial w}{\partial x_i} \end{array} \right\} \tag{2.28}$$

Table 2.1 Beam Element Stiffness Matrix

KA												$dx1$
0	KBz											$dy1$
0	0	KBy										$dz1$
0	0	0	KC		Symmetric							$\theta x1$
0	-KDz	0	0	KEz								$\theta y1$
0	0	KDy	0	0	KEy							$\theta z1$
-KA	0	0	0	0	0	KA						$dx2$
0	-KBz	0	0	KDz	0	0	KBz					$dy2$
0	0	-KBy	0	0	-KDy	0	0	KBy				$dz2$
0	0	0	-KC	0	0	0	0	0	KC			$\theta x2$
0	-KDz	0	0	KFz	0	0	KDz	0	0	KEz		$\theta y2$
0	0	-KDy	0	0	KFy	0	0	-KDy	0	0	KEy	$\theta z2$

where

$$KA = \frac{AE}{L}$$

$$KBy = \frac{12EI_y}{L^3} \qquad KBz = \frac{12EI_z}{L^3}$$

$$KC = \frac{GJ}{L}$$

$$KD_y = \frac{6EI_y}{L^2} \qquad KBz = \frac{6EI_z}{L^2}$$

$$KE_y = \frac{4EI_y}{L} \qquad KEz = \frac{4EI_z}{L}$$

$$KF_y = \frac{2EI_y}{L} \qquad KFz = \frac{2EI_z}{L}$$

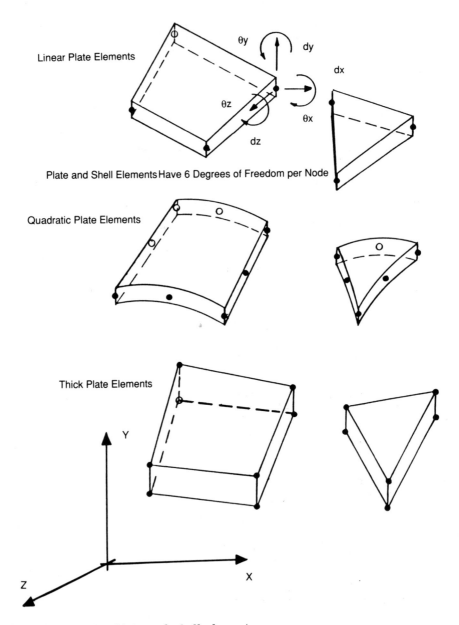

Linear Plate Elements

θy dy dx

θz θx

dz

Plate and Shell Elements Have 6 Degrees of Freedom per Node

Quadratic Plate Elements

Thick Plate Elements

Y

Z X

Figure 2.12 Plate and shell elements.

The B matrix can be given by

$$[B] = \begin{Bmatrix} -\dfrac{\partial^2[N_i]}{\partial x^2} \\[2mm] -\dfrac{\partial^2[N_i]}{\partial y^2} \\[2mm] 2\,\dfrac{\partial[N_i^2]}{\partial x \partial y} \end{Bmatrix} \tag{2.29}$$

The elasticity matrix [D] for an isotropic plate is

$$[D] = \frac{Et^2}{12(1-\mu^2)} \begin{bmatrix} 1 & \mu & 0 \\ \mu & 1 & 0 \\ 0 & 0 & (1-\mu)/2 \end{bmatrix} \tag{2.30}$$

where t is the thickness of the plate.

Quadrilateral plate elements have problems in which the four corner nodes do not form a plane. Most programs will give either a warning or fatal error message when the amount of warping exceeds a small, predetermined tolerance. In cases where warping cannot be avoided, two triangular elements can be substituted for one quadrilateral element.

2.8 ELEMENT INTEGRATION ORDER

Element integration order refers to the number and type of equations describing the displacement field of the element. For isoparametric elements, integration order is tied to the number of nodes. Elements can be broken down into three broad categories.

1. *Linear* (LD), giving a linear variation in displacement within each element and constant strain (Figure 2.13).
2. *Linear with added displacement shapes* (LDADS), giving pseudoquadratic properties in bending while using the same number of degrees of freedom as the linear element (Figure 2.14).
3. *Quadratic*, giving a quadratic or second-order variation in displacement within the element and a linear variation in strain (Figure 2.15).

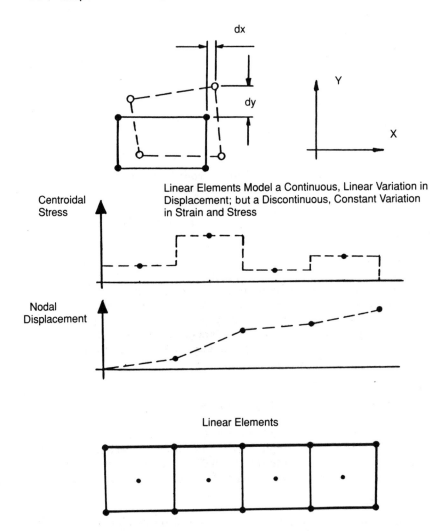

Figure 2.13 Linear elements: stress and displacement variation.

2.8.1 Linear Elements

The element derivation given in Section 2.5 is for a linear two-dimensional element. There is a linear variation in displacement within the element due to the fact that displacements are calculated only

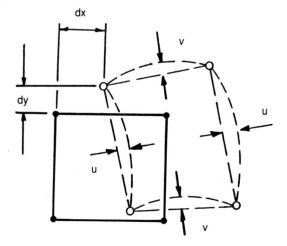

These Elements Allow Bending of Their Sides to Compensate for
the Inherent Overstiffness of Linear Elements in Bending

Figure 2.14 Linear element with added displacement modes.

at the vertices. The first derivative of displacement, strain, is
therefore assumed to be constant over the element. There is continuity of displacement between elements along the boundaries.
There is not continuity of strain at the element boundaries, however. The strain varies across the global model as a series of
step functions (Figure 2.13).

2.8.2 Linear with Added Displacement Shapes

The linear element with added displacement shapes looks like the
standard linear element but has additional degrees of freedom not
associated with any node. These are called nodeless variables or
incompatible bending modes. These additional displacement shapes
allow the sides of the element to bend as shown in Figure 2.14.
For a two-dimensional element, this adds two additional degrees
of freedom to the element stiffness matrix for a total of 10. The
B matrix is modified to accommodate the additional degrees of freedom. The stiffness matrix is initially 10 × 10 but is condensed
down to 8 × 8 by solving for the additional degrees of freedom
in terms of the remaining eight. This type of element has the

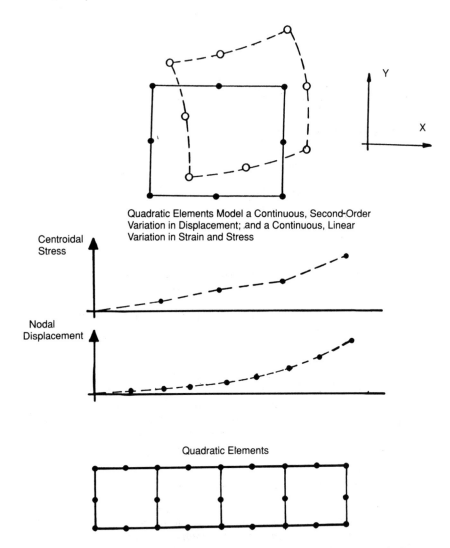

Figure 2.15 Quadratic element stress and displacement variation.

disadvantage of possibly not being continuous along its boundaries. It may fail a "patch test." In practice, this is found to
occur only in rare cases most likely due to severe element distortion. These elements are subparametric because they use more

terms to interpolate the displacement field than they do to interpolate the coordinates [1]. A more complete description of these
elements can be found in [3] or [4].

2.8.3 Quadratic Elements

Quadratic or second-order elements interpolate displacements at
midside locations as well as at their vertices (Figure 2.15). These
elements have the advantage of giving a second-order variation in
displacement across the element and, in turn, allow for a linear
variation in strain across the element. Therefore, continuity of
strain (and stress) is preserved at the element boundaries. This
is especially important in areas of high-stress gradients. The
second advantage that these elements have is the ability to represent directly a curved boundary. With linear elements, a curved
boundary must be represented as a series of straight-line segments.
The shape functions for a two-dimensional quadratic element are

$N_1 = 1/4(1-s)(1-t)(-s-t-1)$

$N_2 = 1/2(1-s)(1-t^2)$

$N_3 = 1/4(1-s)(1+t)(-s+t-1)$

$N_4 = 1/2(1-s^2)(1+t)$

$N_5 = 1/4(1+s)(1+t)(s+t-1)$ $\qquad (2.31)$

$N_6 = 1/2(1+s)(1-t^2)$

$N_7 = 1/4(1+s)(1-t)(s-t-1)$

$N_8 = 1/2(1-s^2)(1-t)$

The derivation of the stiffness matrix follows basically the same
procedure as outlined in Section 2.5.

2.9 DYNAMIC ANALYSIS

Finite element dynamic calculations require generation of individual element stiffness and mass matrices that are combined together
to form a global set of equations of motion.

$$[M] \{\ddot{X}\} + [C] \{\dot{X}\} + [K] \{X\} = \{F\} \qquad (2.32)$$

Typically, a separate damping matrix is not formed but damping
multipliers are added to the mass and stiffness matrices. For undamped, free vibrations, i.e., the eigenvalue problem, the set of
equations becomes

$$[M] \{\ddot{X}\} + [K] \{X\} = 0 \tag{2.33}$$

Assuming sinusoidal motion, we get

$$\{X(t)\} = \{X_0\} \sin \omega_n t \tag{2.34}$$

and

$$\{\ddot{X}(t)\} = -\omega_n^2 \{X_0\} \sin \omega_n t \tag{2.35}$$

therefore,

$$\{-\omega_n^2 [M] + [K]\} \{X_0\} \sin \omega_n t = 0 \tag{2.36}$$

where $\{X_0\}$ is an eigenvector (mode shape) associated with a unique eigenvalue (natural frequency) ω_n.

In the case of harmonic or transient forced response, the force time history and damping values must also be specified. Damping can be included as a separate matrix or as multipliers used with the stiffness and mass matrices. When a separate damping matrix is used, it is out of phase with the stiffness and mass matrices and the damping terms must be handled with complex arithmetic, i.e.,

$$\{-\omega^2 [M] + [K]\} \cos \omega t + \{-\omega[C]\} \sin \omega t = F(t) \tag{2.37}$$

For harmonic motion

$$F(t) = F_0(\cos \omega t + \phi)$$

For transient motion

$$F(t) = F_0(t+t_0)$$

The damping multipliers α and β may be used with the mass and stiffness matrices as

$$\{-\omega^2\alpha[M] + \beta[k]\} \sin \omega t = F(t) \tag{2.38}$$

The α and β multipliers give a damping ratio (C/C_c) as a function of frequency as shown in Figure 2.16.

2.10 SUBSTRUCTURING

Substructuring is a technique used when there is repetitive geometry within a model. A typical example would be a gear with uniform teeth. The generation of stiffness matrices for a structure like this is inefficient because identical element matrices would be

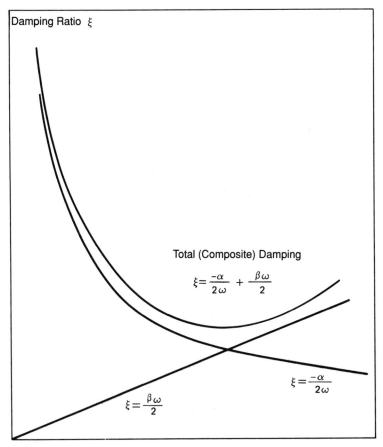

Figure 2.16 α and β damping parameters.

generated in different parts of the structure. With substructuring, element matrices for the portion of the structure to be repeated are generated and internal degrees of freedom are solved for, with degrees of freedom retained along the boundaries for connection with other substructures and other elements. Substructuring is also used in dynamic analyses in which the time-dependent problem must be repeatedly resolved at different time increments to keep the active degrees of freedom in the problem to a minimum.

In the normal solution of a set of matrix stiffness equations, the stiffness matrix is reduced (triangularized), i.e., modified until all terms below the diagonal are 0. The example of Section 2.7 is repeated here.

$$
\begin{bmatrix}
k_{11} & k_{12} & k_{13} & k_{14} & k_{15} \\
k_{21} & k_{22} & k_{23} & k_{24} & k_{25} \\
k_{31} & k_{32} & k_{33} & k_{34} & k_{35} \\
k_{41} & k_{42} & k_{43} & k_{44} & k_{45} \\
k_{51} & k_{52} & k_{53} & k_{54} & k_{55}
\end{bmatrix}
\begin{Bmatrix}
d_1 \\ d_2 \\ d_3 \\ d_4 \\ d_5
\end{Bmatrix}
=
\begin{Bmatrix}
F_1 \\ F_2 \\ F_3 \\ F_4 \\ F_5
\end{Bmatrix}
\tag{2.39}
$$

It is reduced to look like this

$$
\begin{bmatrix}
k'_{11} & k'_{12} & k'_{13} & k'_{14} & k'_{15} \\
0 & k'_{22} & k'_{23} & k'_{24} & k'_{25} \\
0 & 0 & k'_{33} & k'_{34} & k'_{35} \\
0 & 0 & 0 & k'_{44} & k'_{45} \\
0 & 0 & 0 & 0 & k'_{55}
\end{bmatrix}
\begin{Bmatrix}
d_1 \\ d_2 \\ d_3 \\ d_4 \\ d_5
\end{Bmatrix}
=
\begin{Bmatrix}
F'_1 \\ F'_2 \\ F'_3 \\ F'_4 \\ F'_5
\end{Bmatrix}
\tag{2.40}
$$

The last equation in this set now has only one unknown d_5 and can be explicitly solved for. The second to last equation can be solved using the previously obtained value of d_5 and solving for d_4. The process continues back through the set of equations until all the unknowns $\{d\}$ are solved for. This two-step process is known as forward-elimination and back-substitution.

In forming a substructure, the set of equations is reordered to place the retained (boundary) degrees of freedom last and the to be eliminated (internal) degrees of freedom first. These degrees of freedom are referred to as master and slave degrees of freedom, respectively. Slave degrees of freedom are then solved in terms of the master degrees of freedom. Actual displacements for the slave degrees of freedom can be obtained after the displacements of the master degrees of freedom are explicitly solved for.

In the example (2.38), if degrees of freedom 1 and 5 are masters and degrees of freedom 2, 3, and 4 are slaves, then the set of equations can be partitioned as follows:

$$\begin{bmatrix} k_{21} & k_{22} & k_{23} & k_{24} & k_{25} \\ k_{31} & k_{32} & k_{33} & k_{34} & k_{35} \\ k_{41} & k_{42} & k_{43} & k_{44} & k_{45} \\ k_{11} & k_{12} & k_{13} & k_{14} & k_{15} \\ k_{51} & k_{52} & k_{53} & k_{54} & k_{55} \end{bmatrix} \begin{Bmatrix} d_2 \\ d_3 \\ d_4 \\ d_1 \\ d_5 \end{Bmatrix} = \begin{Bmatrix} F_2 \\ F_3 \\ F_4 \\ F_1 \\ F_5 \end{Bmatrix} \qquad (2.41)$$

The forward-elimination process is carried out through the slave degrees of freedom but stops short of the master degrees of freedom. At this point, the reduced set of equations looks like

$$\begin{bmatrix} k'_{21} & k'_{22} & k'_{23} & k'_{24} & k'_{25} \\ 0 & k'_{32} & k'_{33} & k'_{34} & k'_{35} \\ 0 & 0 & k'_{43} & k'_{44} & k'_{45} \\ k'_{11} & k'_{12} & k'_{13} & k'_{14} & k'_{15} \\ k'_{51} & k'_{52} & k'_{53} & k'_{54} & k'_{55} \end{bmatrix} \begin{Bmatrix} d_2 \\ d_3 \\ d_4 \\ d_1 \\ d_5 \end{Bmatrix} = \begin{Bmatrix} F'_2 \\ F'_3 \\ F'_4 \\ F'_1 \\ F'_5 \end{Bmatrix} \qquad (2.42)$$

The reduced set of equations is now in two parts

$$\begin{bmatrix} k_{ss} & k_{sm} \\ k_{ms} & k_{mm} \end{bmatrix} \times \begin{Bmatrix} d_s \\ d_m \end{Bmatrix} = \begin{Bmatrix} F_s \\ F_m \end{Bmatrix} \qquad (2.43)$$

The subset of the equations representing the master degrees of freedom is the substructure model

$$[k_{mm}] \{d_m\} = \{F_m\} \qquad (2.44)$$

and the $[k_{mm}]$ matrix can be used like an element matrix, representing the equivalent stiffness of the entire structure. A substructure is also known as a superelement.

The matrix $[k_{ss}]$ is the back-substitution matrix and is held aside until the global problem is solved and the displacements of the master degrees of freedom $\{d_m\}$ are available. At this point, the back-substitution process can be completed for the slave degrees of freedom and their displacements $\{d_s\}$. Once the slave degrees of freedom are solved for, the element stresses can be obtained in the normal manner.

When this process is applied to static analysis, it is known as static condensation. For dynamic problems, a similar procedure is followed to reduce both the stiffness and mass matrices. It is called dynamic condensation. One of the most popular methods of dynamic condensation is Guyan reduction [5] that reduces the stiffness matrix independently of the mass matrix using static condensation techniques. For mass condensation, Guyan reduction redistributes the mass according to the reduced stiffness matrix. The mass is redistributed with more mass being placed in areas of high stiffness, as these areas tend to behave as rigid bodies within the structure.

In the following simple example, shown in Figure 2.17, nodes 5, 6, and 7 are connected by spring elements. Each node has only one degree of freedom and mass is assigned to each node. Node 6 is to be eliminated. The equivalent stiffness of k'_{5-7} is simply

$$k'_{5-7} = \frac{1}{\dfrac{1}{k_{5-6}} + \dfrac{1}{k_{6-7}}} \tag{2.45}$$

Equivalent Stiffness Is First Determined and Resolution of the Removed Mass Is Based on the Relative Stiffness of the Two Original Springs

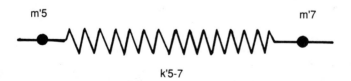

Figure 2.17 Stiffness and mass resolution.

The mass of node 6 will be distributed to nodes 5 and 7 according to the relative stiffness of k_{5-6} and k_{6-7}.

$$m_5' = m_5 + m_6 \frac{k_{5-6}}{k_{6-7}'} \tag{2.46}$$

2.11 GENERALIZED FIELD PROBLEMS

The finite-element method can be used to solve a variety of field problems in addition to structural stress and vibrations. Problems in heat transfer, fluid flow, and electromagnetics can be solved using analogous relationships. These field problems behave according to the Laplace and Poisson equations.

The general form of these equations can be given as

$$\frac{\partial}{\partial x}\left(k_x \frac{\partial \phi}{\partial x}\right) + \frac{\partial}{\partial x}\left(k_y \frac{\partial \phi}{\partial y}\right) + \frac{\partial}{\partial z}\left(k_z \frac{\partial \phi}{\partial z}\right) + Q = 0 \tag{2.47}$$

The analogous parameters for various field problems are given in Table 2.2. The general formulation of the matrix equations is based on the minimization of the volume integral

$$X = \frac{1}{2} \int k_x \left(\frac{\partial \phi}{\partial x}\right)^2 + k_y \left(\frac{\partial \phi}{\partial y}\right)^2 + k_z \left(\frac{\partial \phi}{\partial z}\right)^2 - Q\phi \ dxdydz \tag{2.48}$$

Table 2.2 Analogous Parameters in Field Problems

Problem	Nodal variable	Forcing	Matrix
Static displacement	Displacement	Force	Stiffness
Heat conduction	Temperature	Heat flux	Conductivity
Ideal fluid flow	Flow potential	Fluid flux	Geometry-based
Magnetic	Magnetic potential	Magnetic field	Magnetic permeability

To minimize the above integral over the element domain, the derivative with respect to the nodal variables ϕ_i is taken

$$\frac{\partial X}{\partial \phi} = \int k_x \frac{\partial \phi}{\partial x} \frac{\partial}{\partial \phi_i} \frac{\partial \phi}{\partial x} + k_y \frac{\partial \phi}{\partial y} \frac{\partial}{\partial \phi_i} \frac{\partial \phi}{\partial y}$$

$$+ k_z \frac{\partial \phi}{\partial z} \frac{\partial}{\partial \phi_i} \frac{\partial \phi}{\partial z} - Q \frac{\partial \phi}{\partial \phi_i} \, dxdydz \qquad (2.49)$$

$$+ \int \frac{\partial \phi}{\partial \phi_i} q \quad \frac{\partial \phi}{\partial \phi_i} a\phi \quad dS$$

where the second integral is the integral around the boundary of the domain. It is assumed that the nodal variables ϕ_i can be related to any value of ϕ within the element by the shape functions such that

$$\phi = [N]\{\phi_i\} \qquad (2.50)$$

and that the derivatives of the variables can be obtained from the derivatives of the shape functions and the nodal variables ϕ_i.

The equivalent "stiffness" matrix is obtained by integrating equations (2.47) over the element volume

$$k_{i,j} = \int k_x \frac{\partial N_i}{\partial x} \frac{\partial N_j}{\partial x} + k_y \frac{\partial N_i}{\partial y} \frac{\partial N_j}{\partial y} + k_z \frac{\partial N_i}{\partial z} \frac{\partial N_j}{\partial z} \, dxdydz \qquad (2.51)$$

The equivalent "force" vector is given by

$$F_i = - \int Q \, N_i dV + q \, N_i \, dS + \int [N] \, a \, N_i \, dS \, \{\phi\} \qquad (2.52)$$

where the volume term represents internal flux terms such as heat generation, etc.

The solution of the variety of generalized field problems is treated in a number of sources such as [1, 6, 7, 9] for the general cases. Fluid problems are specifically discussed in [10−12]. A discussion of magnetic applications is found in [13].

2.12 SUMMARY

This chapter represents only a brief overview into the fundamentals of the finite-element method. Discussion of the fundamentals has been kept brief because there are a number of excellent sources on fundamentals referenced in this chapter, and the purpose of this book is to discuss finite element modeling fundamentals. In order to efficiently apply the finite element method, however, it is necessary to have a knowledge of the mathematical principles. For a more in-depth discussion of any of the chapter topics, the reader is referred to one of the above-mentioned sources.

REFERENCES

1. Zienkiewicz, O. C., *The Finite Element Method in Engineering Science*, 2nd ed., McGraw-Hill, London, 1971.
2. Irons, B. M., "A Frontal Solution Program for Finite Element Analysis," *International Journal for Numerical Methods in Engineering*, Vol. 2, No. 1, Jan. 1970, pp. 5–23.
3. Cook, R. D., *Concepts and Applications of Finite Element Analysis*, Wiley, New York, 1974.
4. Gallagher, R. H., *Finite Element Analysis Fundamentals*, Prentice-Hall, Englewood Cliffs, N.J., 1975
5. Guyan, R. J., "Reduction of Stiffness and Mass Matrices," *AIAA Journal*, Vol. 3, No. 2, Feb. 1965.
6. Zienkiewicz, O. C. and Y. K. Cheung, "Finite Elements in the Solution of Field Problems," *The Engineer*, Sept. 1965, pp. 507–510.
7. Heubner, K. H., *The Finite Method for Engineers*, Wiley, New York, 1975.
8. Bathe, K. J. and E. L. Wilson, *Numerical Methods in Finite Element Analysis*, Prentice-Hall, Englewood Cliffs, N.J., 1976.
9. Huston, R. L. and C. E. Passerello, *Finite Element Methods: An Introduction*, Marcel Dekker, New York, 1984.
10. Gallagher, R. H., J. T. Oden, O. C. Zienkiewicz, and C. Taylor, *Finite Element Methods in Fluids*, Vol. 1 and 2, Wiley, New York, 1975.
11. Martin, H. C., "Finite Element Analysis of Fluid Flows," Proceedings of Second Conference on Matrix Methods in Structural Mechanics, Wright Patterson Air Force Base, Dayton, Ohio, 1968.
12. deVries, G. and D. H. Norrie, "The Application of the Finite-Element Technique to Potential Flow Problems," *Journal of Applied Mechanics*, Transaction of ASME, Series E, Vol. 38, 1971, pp. 798–802.

13. Kohnke, P. C. and J. A. Swanson, "Thermo-Electric Finite Elements," Proceedings of International Conference on Numerical Methods in Electrical and Magnetic Field Problems, June 1–4, 1976, Santa Margherita Liqure, Italy.

3

BASIC INPUT DATA

The input to various finite element programs may appear to be different; however, all programs require the same basic input data. In forming a finite element model, the engineer must be knowledgable in the following three areas:

1. The basic behavior of the structure to be modeled.
2. The required input data for the program that he will use.
3. An understanding of modeling techniques to make the most cost-effective model.

Before the engineer can understand cost-effective modeling techniques, he must be familiar with the basic finite element input data requirements.

Every finite element program requires certain generic input data. These data include

1. Definition of the structure's geometry by node and element data.
2. Specification of material properties.

   3. Specification of displacement constraints.
   4. Specification of applied forces.

For cases other than static stress analysis such as thermal, fluid flow calculations, etc., the parameters such as displacement constraints and applied forces are replaced by their analogous variables, such as nodal temperature, heat flux, etc. For the present purposes, static stress analysis will be used as the default case.

3.1 GEOMETRY DEFINITION

The structure's geometry is specified in terms of node and element input. The nodes (or nodal points) are defined in terms of their coordinates and elements are defined by the nodes that bound them.

   Nodes are input by giving their node number and coordinates. Nodes may be specified in any of several coordinate systems: global Cartesian, cylindrical, or local. The available coordinate system options depend on the individual program. It should be noted that the finite element method is nondimensional and all dimensions and values must be checked by the user for consistency. Typically in the English system, inches are used as the linear measure, and in the CGS system, millimeters are used. However, any unit of measure may be used so long as it is consistent with the units of force, Young's modulus, and density. The origin of the coordinate system is entirely arbitrary.

   Figure 3.1 shows a pattern of nodes that will be used to form a simple two-dimensional beam. Finite element models can be classified as either two-dimensional, where all forces and displacements act in plane, or three-dimensional, where full three-dimensional behavior is anticipated. Figure 3.1 shows a two-dimensional nodal pattern so that only x and y coordinates need to be input; z is assumed to be 0.

   Elements are defined and input according to their boundary nodes. Element numbering for some programs must go in a counterclockwise direction. The total number of nodes defining an element depends on the element type. Some basic element types are shown in Figure 3.2 and are described in more detail in Chapter 7. Figure 3.3 shows the pattern of nodes of Figure 3.1 formed into elements that describe the beam. This beam model is constructed from quadrilateral, four-node, two-dimensional elements. Figure 3.4 shows the same beam model generated by triangular, three-

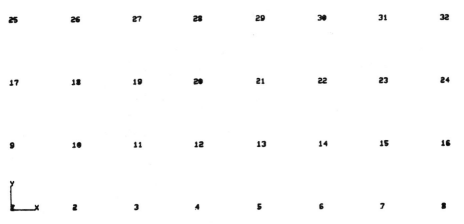

Figure 3.1 Node pattern: Simple beam.

node, two-dimensional elements. A simple element input list would consist of the element number, possibly an element-type identifier and the list of three or four node numbers describing the element. Figure 3.5 shows a beam generated from eight-node, three-dimensional elements. In this simple case, the minimum of two planes of nodes is required.

3.2 MATERIAL PROPERTIES

For static stress analysis, the only material properties required are Young's modulus and Poisson's ratio because only the stiffness of the structure needs to be calculated. For dynamic cases, the material mass density must also be input. Care must be taken to insure that the units used for mass density are consistent with the units of length, time, acceleration, and force. There will not be one specific set of units specified in any finite element program user's manual and there is no error checking for units within the programs. Typical units for the English system might be

Length, in.
Time, sec
Force, lbf
Mass, lbf/in./sec^2

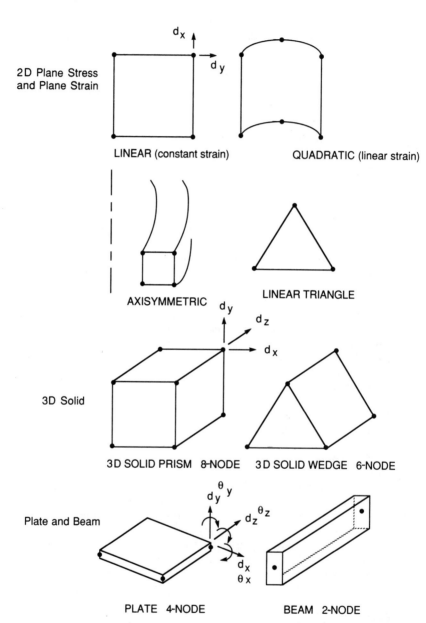

Figure 3.2 Basic element types.

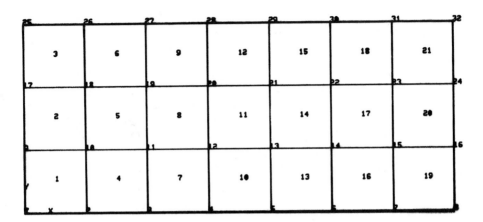

Figure 3.3 Element pattern: Simple beam.

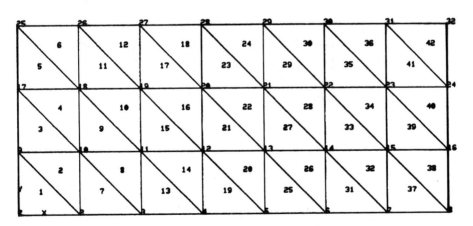

Same Number of Nodes and Degrees of Freedom
Each Quadrilateral Element is Replaced by Two Triangles

Figure 3.4 Triangular element pattern: Simple beam.

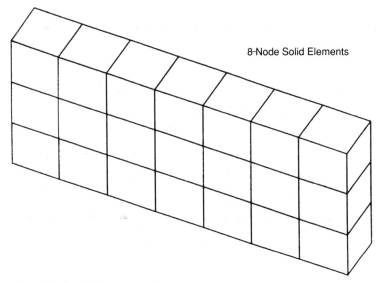

8-Node Solid Elements

Same Number of Elements as the 2D Quadrilateral Case but with
Twice as Many Nodes and Three Times as Many Degrees of Freedom

Figure 3.5 Three-dimensional element pattern: Simple beam, eight-node solid elements.

Certain element types, such as plate and beam elements, require specialized material and cross-sectional property data such as area and area moments of inertia. For example,

    Plate element
      Young's modulus
      Poisson's ratio
      Density (if dynamics are involved)
      Shear modulus (optional)
      Thickness
    Beam element
      Young's modulus
      Poisson's ratio
      Density (if dynamics are involved)
      Shear modulus (optional)
      Cross-sectional area(s)
      Area moments of inertia about two local axes and the orientation of those axes with respect to a global set of axes

3.3 DISPLACEMENT CONSTRAINTS

Displacements must be constrained at one or more points in the model. All degrees of freedom must be constrained at least one point to prevent rigid body motion of the model. In cases where it may be possible to input balanced, external forces on the model, i.e., where there is no net external force or moment, it is still advisable to constrain at least one point. Computer roundoff and truncation errors can lead to a small, but finite net unbalanced force on the model.

Constrained nodes may have between one and all of its degrees of freedom constrained. When not all degrees of freedom are constrained, the node may be said to act as though it were on rollers. For plate and beam elements where there are six degrees of freedom per node, constraints on the displacement degrees of freedom without constraints on the rotational degrees of freedom would represent a pinned connection. Figure 3.6 shows common methods of nodal constraints. In specifying nodal constraints, the node number is given followed by a code to identify the various degrees of freedom and a code to specify whether they are to be constrained or left free. For example, in the case of the simple beam, nodes 1 through 4 should be constrained in the x direction and node 1 should also be constrained in the y direction. This arrangement allows the beam to deform in the y direction, due to Poisson's effect. Note that this constraint arrangement would not give the same result as the case where all degrees of freedom at nodes 1 through 4 were completely constrained. In the first case, the constraints will represent half of a uniform free-free beam. In the second cases, the displacement constraints model a clamped-free beam. Typically, constrained nodes are specified as having zero displacement; however, nonzero displacements can also be specified. These nonzero displacements are used in cases of thermal expansion, etc.

A common problem in specifying displacement constraints is overconstraining the model by specifying too many constraints. This can lead to locally high stresses and unrealistic behavior. In specifying displacement constraints, only the minimum number of constraints should be used.

3.4 APPLIED FORCES

Regardless of how forces are input to the finite-element program, all forces are converted into point loads applied at the nodal points.

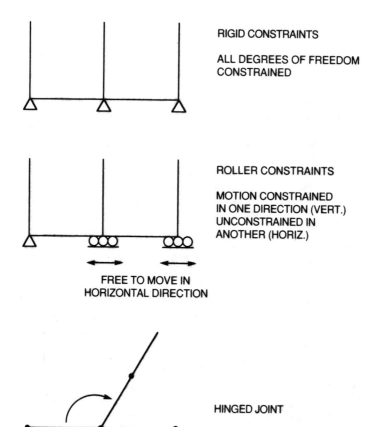

RIGID CONSTRAINTS

ALL DEGREES OF FREEDOM
CONSTRAINED

ROLLER CONSTRAINTS

MOTION CONSTRAINED
IN ONE DIRECTION (VERT.)
UNCONSTRAINED IN
ANOTHER (HORIZ.)

FREE TO MOVE IN
HORIZONTAL DIRECTION

HINGED JOINT

BEAM OR PLATE ELEMENT
CONNECTED TO SOLID
ELEMENTS

Figure 3.6 Common methods of nodal constraints.

Applied forces can be generally categorized into the following three groups:

1. Direct nodal forces.
2. Distributed, pressure-type forces.
3. Body forces.

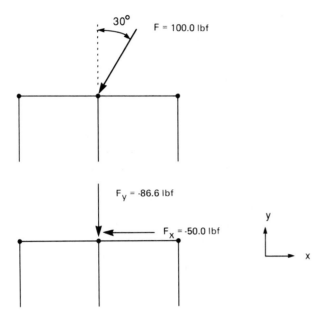

Figure 3.7 Nodal force.

3.4.1 Direct Nodal Forces

Nodal forces are specified according to the degree of freedom so that each degree of freedom of a node may have a different force amplitude. The direction of force is specified by the sign of the force amplitude relative to the global coordinate system. Forces not coincident with the global axes are represented by the vector sum of the individual force components. Figure 3.7 shows a compressive force applied at a node with an amplitude of 100.0 and at a direction of 30° from the vertical. This force would be broken down into its x and y components as

$$F = 100.0 \text{ at } 60° = > F_x = 100.0 \cos(60) = -50.0$$

$$F_y = 100.0 \sin(60) = -86.6$$

A fixed format force input for the force might be

Node no.	F_x	F_y
102	-50.0	-86.6

3.4.2 Distributed Forces

Distributed forces such as pressure are almost always input to
the program by specifying a range of elements over which the
pressure acts, the direction and amplitude of the pressure. The
program, in turn, breaks down the pressure into the appropriate
nodal forces. When pressure is applied to a surface, the equivalent nodal forces are calculated by the product of the node's share
of the area and pressure over that area. Figure 3.8 shows a simple one-dimensional example of a uniform pressure applied over
two series of nodes, one with uniform pressure and the second
with nonuniform pressure. Note that with the case of uniform
pressure, the end nodes only get half the force of the interior
nodes. Although this appears to apply a nonuniform force to
the model, it does result in the correct pressure distribution.

For linear elements with corner nodes only, the resolution of
applied pressure to nodal forces follows a linear pattern, with
each node receiving its share of force based on the product of
its equivalent area and applied pressure as shown in Figure 3.9.
For quadratic elements, with midside nodes, the force resolution
scheme is more complex, as shown in Figure 3.10. The force resolution for the quadratic elements is not intuitively obvious and
is the result of the integration of pressure over the element surface using the element's integration functions.

Because forces are applied only at the node points, it is impossible to achieve a perfectly distributed or continuously variable force. As force varies from node to node, it is represented
by the element as either a step or linear ramp change. For this
reason, stress results in elements with applied nodal forces will
not yield the best results. Internal forces and stresses will
"even out" within a few layers of elements so that the stresses
will be of normal accuracy within a short distance of the nodal
forces. The general rule of thumb is to use a radius of three
elements away from the element with the applied force to find
good stress results. In areas where each node is carrying the
same force, this distance should be less because it is the step
changes in applied force that locally affect the stress results.
In cases where it is important to calculate stresses near a surface with an applied load, several layers of small elements are
recommended.

3.4.3 Body Forces

Another class of applied forces is body forces. Included are
forces such as centrifugal or magnetic forces that act throughout

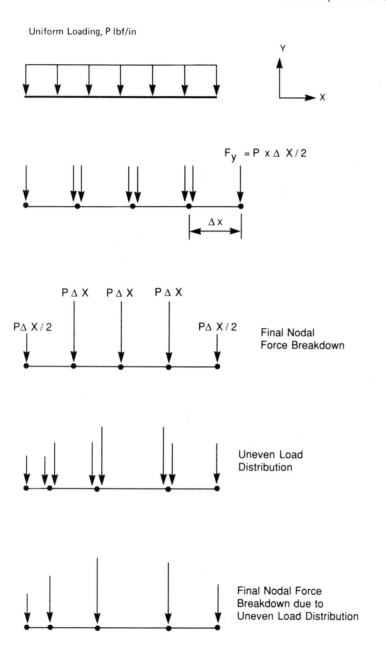

Figure 3.8 Resolution of pressure loading into nodal forces.

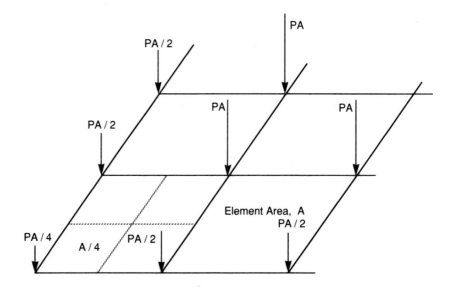

Figure 3.9 Resolution of uniform pressure over linear elements.

the structure and are applied to every element. Often, these forces are dependent on element properties such as mass density or magnetic permeability. Body forces are always applied by specifying the global input variable, such as rotational speed for centrifugal forces, and allowing the program to calculate nodal forces on the element level.

In the case of centrifugal forces, the input parameters are the rotational speed, axis of rotation, and absolute value of the coordinates. In this case, the origin of the coordinate system is not arbitrary as the magnitude of the centrifugal force is directly proportional to the radial distance from the axis of rotation, i.e.,

$$F_c = mr\omega^2 \tag{3.1}$$

3.5 AUTOMATIC MESH GENERATION

Input data are never entered one node at a time or one element at a time. Even in the pre-CAD days before interactive computer graphics, autogeneration and repeat commands were available in every commercial finite element program to minimize the amount of input data required. With the advent of true three-dimensional

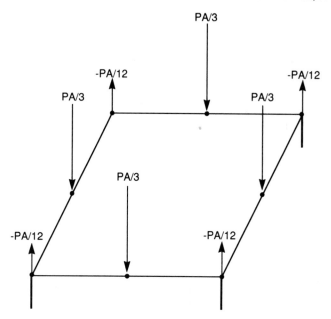

Figure 3.10 Resolution of uniform pressure over quadratic elements.

solid modeling and sophisticated computer graphics systems, the input of geometry data may hardly resemble the node and element system. Areas and volumes may be defined, manipulated, and "meshed" with node and element patterns of selected densities. The fundamental parameters of nodes, elements, nodal displacement constraints, and applied nodal forces still apply, however. The automatic mesh generation programs are only a front end to the finite element program and generate a file containing all nodes and elements explicitly defined.

3.6 DATA INPUT FORMAT

The actual format of input data varies from program to program. Some programs use a fixed or free format card image system, whereas other programs are intended for interactive data input

Table 3.1 Examples of Two Data Input Formats

Parameter	ANSYS	MSC/NASTRAN
Node definition		

Node 1 at x = 1.00
     y = 0.00
     z = 2.00
	N, 1, 1, 0, 2	GRID, 1,, 1, 0, 2

Two-dimensional element with nodes 1, 2, 6, 5	E, 1, 2, 6, 5	CQUAD4, 1, 1, 2, 6, 5
Displacement constraint at node 1 with all three D.O.F. constrained	D, 1, ALL	SPC1, 1, 123456, 1

and rely heavily on graphics, incorporating data editing commands along with data input commands. A comparison of two of the more popular programs, ANSYS and MSC/NASTRAN, shows the different systems for basic data entry (see Table 3.1).

3.7 SUMMARY

A study of modeling principles must be preceded by a knowledge of basic finite-element input data, especially in the case of the novice finite element user. Modeling is actually the optimization of basic input data for a given problem; therefore, a knowledge of what data needs to be optimized is essential. As the data input procedures for the various finite element programs grow more sophisticated, they become more removed from the basic input data. Explicit node and element definition is being replaced by line, area, and volume definition and automatic mesh generation. In using the more sophisticated autogeneration routines, the engineer must keep in mind the basics of model building in order to develop a cost-effective model and to ensure that the autogeneration routines give the appropriate model.

4

PROBLEM DEFINITION

The previous two chapters have given theoretical background information and generic data input requirements for finite element analysis. The objective of this chapter is to give some guidelines for combining the user's knowledge of the structure or machine component to be analyzed, together with the data input requirements to form the most cost-effective model that will yield acceptable results. The definition of modeling given here is the *process of examining an engineering problem and developing the most cost-effective set of input data that will help solve the problem.*

To put Chapters 2—4 in perspective, an analogy can be made to teaching someone to drive an automobile. The theory discussed in Chapter 2 is analogous to a discussion of the internal combustion engine and an explanation of how a transmission and brakes work. These are the under-the-hood details that are not necessarily required to operate the automobile, but are recommended for a basic understanding of it. Chapter 3 provides a set of generic data input requirements that are applicable to almost any finite element program. This is analogous to basic automobile operation principles. All automobiles are operated in basically the same manner. For example, the brake is operated by a pedal on the floor, etc. In this current chapter, guidelines for the intelligent applications of finite element analysis will be given so that the finite element "automobile" travels to its destination in the most direct manner and with the best "mileage." In driving, no one would drive with one foot on the accelerator and one foot on the brake for any length of time; however, with finite element applications, it is possible to develop a model that will consume "gallons" of computer time without providing the required result.

The optimum model development procedure can be generally broken down into three steps

Problem definition.
Free-body diagram.
Finite-element model.

The generation of the actual finite element gridwork is the most difficult part of the modeling process and this is what most engineers think of when they hear the word "modeling." This task can be simplified by breaking down the entire modeling process into discrete steps so that the process of gridwork layout can be isolated and handled in an organized manner. Steps 1 and 2, problem definition and the free-body diagram, will be covered in this chapter and finite element model generation in Chapter 5.

Six examples will be used throughout this, and subsequent, chapters to illustrate this procedure: a plate with a circular hole in uniaxial tension, a notched block, a pressure vessel, a welded crank, two gears in contact, and a turbine blade.

4.1 PROBLEM DEFINITION

The problem definition step requires the most nonfinite element experience and should take place before the engineer sits down at the computer terminal. The problem definition step requires a knowledge of the structure to be analyzed and its anticipated behavior. In this step, a number of questions must be answered and the engineer needs to have, at least, a mental checklist of items to be determined and quantified. This list should include, at a minimum, the

Objectives of the analysis.
Exact geometric description of the structure.
Forces acting on the structure.
Displacement constraints acting on the structure.

In determining the *objectives of the analysis*, the following questions must be answered:

1. What is the specific engineering problem being addressed? It may be a failure of an existing structure, the likelihood of failure of a new prototype structure, the amount of deflection of the structure during operation, or the amount of vibration produced or transmitted by the structure during operation. In many cases, there may be more than one objective to the analysis. The time to determine all the information that will be required from the analysis is at this, first step.

2. What level of accuracy is expected or required from the analysis? Is the purpose of the analysis to provide "ballpark" numbers and general guidance for design or is the analysis of a critical component that must be precisely designed? What will the consequences be if the component or structure fails in the future?

3. What parameters will affect the performance of this structure? Such parameters include static loading and stress, vibration and/or resonance, thermal effects, etc. Does the structure behave in a linear manner or will nonlinear effects such as creep or buckling have to be taken into account? If there are nonlinear effects that cannot be included in the finite-element analysis, how will they affect the final accuracy of the analysis?

The *exact geometric description* of the component or structure depends not only on design drawings but also on tolerance effects. If an analysis of an existing structure is required, it is important to check the structure against design specifications to make sure that it meets those specifications and to take note of any exceptions. When tolerances and manufacturing techniques are important, such as the dynamic stiffness of a bolted joint, it is necessary to quantify the range of composite tolerances. Consideration should be given to making several parametric calculations to bracket the tolerance effects.

As part of this step, it is necessary to make a decision as to how much of a structure, machine, or machine component to include in the model. Generally, there are a finite number of areas of interest that are determined from previous experience or as a result of engineering judgment. The decision then is how much of the surrounding structure needs to be included in the model. This is largely dependent on which parameters affect the behavior of the structure and how much confidence the engineer has in anticipating, at least qualitatively, those effects. For example, in the case of vibration, it may be necessary to model the entire structure to obtain the dynamic stress at one point because the dynamic response is dependent on the stiffness and mass distribution of the entire structure. The use of symmetry may be considered to reduce the problem size.

Forces acting on the structure can consist of static discrete forces, distributed forces, dynamic transient forces, dynamic harmonic forces, thermal effects, and electromagnetic effects. In this problem definition step, it is necessary to make assumptions as to which forces to consider and which to ignore. This requires engineering judgment and familiarity with structure and its behavior.

Static forces require quantifying force amplitude and location. Quantifying forces is often related to deciding how much of a structure to include in the model. When forces on one component of a structure or machine are transmitted by another component, it may be necessary to include the second component in the model to get the accurate force distribution on the primary component of interest.

Dynamic forces require (1) force amplitude, (2) location, (3) frequency, and (4) phase angle. For both transient and harmonic forces, there may be more than one time history or frequency involved. In these cases, it is necessary to know the time-phasing relationship between the forces. It is also necessary to quantify the damping acting in the structure. Damping is a process of dissipating energy that can consist of three basic mechanisms: (1) material or hysteretic damping due to internal friction in the structure, (2) coulomb or dry friction damping due to sliding between surfaces, and (3) aerodynamic (viscous) damping due to the motion of a structure through a fluid such as an airplane wing or a shock absorber.

Displacement constraints require some assumptions to be made as to a location in the structure, away from the critical areas, that may be assumed to be "rigid." Although there is no point in any real structure that has no deflection, there are points where the engineer intuitively knows that the deflections will be two or three orders of magnitude lower than the deflections in the critical areas. The decision as to how and where to assume displacement constraints is directly related to the decision on how much of the structure is to be modeled and whether or not symmetry can be used to minimize the size of the model. An ideal situation is one of a relatively flexible, isolated component connected to a massive structure by a rigid connection, i.e., a bracket welded to a machine frame.

Displacement constraints can be classified into several broad categories: (1) rigid in which all degrees of freedom are constrained along an entire boundary, i.e., a welded connection. (2) roller in which displacements are constrained in one or two directions and are free to move in the other direction(s). For example, a loosely bolted connection in which displacements can be constrained between two surfaces in the normal direction but are free to move laterally could be modeled with roller constraints. And (3) gap constraints in which a structure is free to deflect a limited amount before running into a rigid constraint in one or more directions.

In the problem definition step, it is necessary to examine the structure and classify the objectives of the analysis, quantify the size of the model, and classify and quantify the applied forces and displacement constraints. These items can be summarized in a checklist given in Table 4.1.

4.2 EXAMPLE CASES

A set of six example cases are discussed. The cases are arranged in order of increasing complexity. This set of six problems will be used in subsequent chapters to illustrate the modeling process. In this problem definition stage, the problem statement is somewhat arbitrary, so for each case, there could be multiple problem statements, each of which could lead to a different model of the same structure.

4.2.1 Example: Plate with Circular Hole

A simple plate, in uniaxial tension, with a circular hole is shown in Figure 4.1. In this, the simplest case, the model will be from a textbook problem rather than a "real world" practical case. An exact solution exists for the stress field that will allow for comparison with finite element results. The solution used for comparison here is taken from Seeley and Smith [1].

In the case of the plate with the circular hole, the checklist of Table 4.1 can be followed to define the analysis.

1. *Objectives* The problem is to solve for the maximum static stress in the plate to within 10% of the true value. Although this is an example problem, it must be assumed that thermal effects can be neglected and only the applied static forces will affect the stresses.

2. *Exact geometric description* Because this is a simplified example problem, it may be assumed that the nominal dimensions given in Figure 4.1 may be accepted as shown. Tolerance effects will not be included. It can be seen that because of the assumed

Table 4.1 Problem Definition Checklist

1. Objectives
   1.1 What is the problem to be solved?
   1.2 What level of accuracy is to be expected?
   1.3 What are the parameters that can affect the behavior of the structure?

2. Exact geometric description
   2.1 Are drawings of the structure available?
   2.2 Does the actual structure conform to the drawings?
   2.3 What are the manufacturing tolerances specified for the structure and what tolerances are actually expected?
   2.4 What unspecified effects of tolerance and fit can affect the behavior of this structure?
   2.5 How much of the structure needs to be included in the finite element model?
   2.6 Can symmetry be used to reduce the size of the model?

3. Forces
   3.1 What types of forces act on the structure: static, dynamic, or thermal?
   3.2 What is the location and magnitude of these forces?
   3.3 For dynamic forces, what are the frequencies and phase angles of the forces and what is the damping in the structure?
   3.4 For thermal cases, are the temperature distributions known or will a thermal finite element analysis be required to obtain the temperatures?

4. Constraints
   4.1 What types of constraints are present: rigid, roller, or gap?
   4.2 At what locations can constraints be assumed to be applied?
   4.3 For gap-type constraints, what is the limit of deflection?

uniform force, only one-quarter of the plate needs to be included in the model. A quarter-symmetric model is adequate to represent the plate because the stress distribution in each of the quadrants will be mirrored about the planes of symmetry. If the force was nonuniform, then a half-symmetric or full-plate model would need to be used.

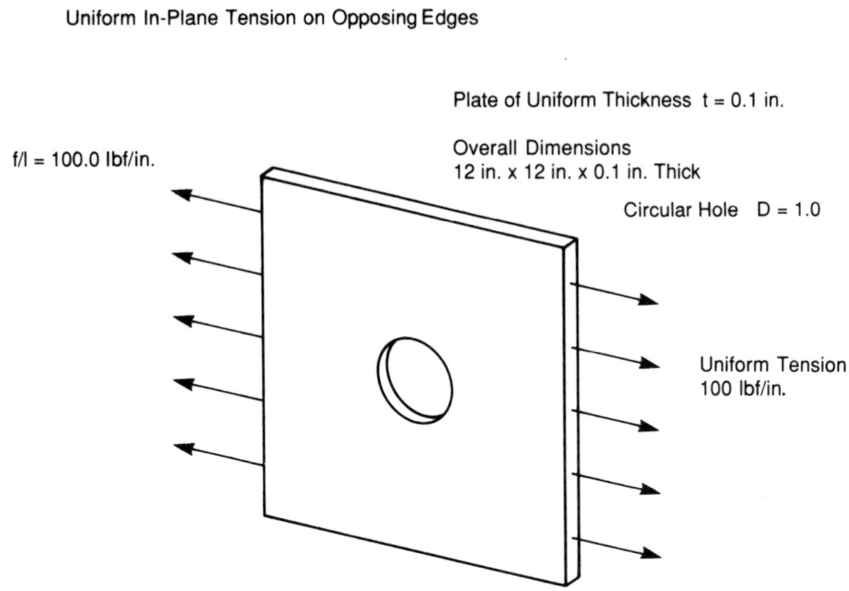

Uniform In-Plane Tension on Opposing Edges

Plate of Uniform Thickness t = 0.1 in.

f/l = 100.0 lbf/in.

Overall Dimensions
12 in. x 12 in. x 0.1 in. Thick

Circular Hole   D = 1.0

Uniform Tension
100 lbf/in.

Figure 4.1 Plate with circular hole: Problem definition.

3. *Forces* The only forces shown in Figure 4.1 are in-plane forces uniformly distributed over the two ends. Therefore, a two-dimensional plane stress model will be sufficient. Obviously, the total forces on each end must be equal for the plate to be in equilibrium.

4. *Constraints* To maintain equilibrium and proper displacements, roller constraints should be specified along the symmetric, cut boundaries. Along the X axis cut boundary, displacements should be constrained in the Y direction, but the surface should be free to move in the X direction. A similar situation holds for the Y axis cut boundary where the X direction should be constrained and the Y direction free to move.

The use of the quarter-symmetric model with force on one face and roller constraints on the symmetric boundaries will eliminate the problem of trying to apply perfectly balanced forces to the finite-element model. Due to computer truncation and round-off error, it is practically impossible to input sets of forces that will balance and not result in a net force.

4.2.2 Example: Notched Block

A simple block with a notch is shown in Figure 4.2. This is, again, a simplistic example but it serves to illustrate the modeling process. Two force cases can be considered, uniform tension and a tensile gradient across the end faces.

A review of the checklist of Table 4.1 might yield the following:

1. *Objectives* The problem is to solve for the maximum static stress in the block. It will be assumed that thermal effects may be neglected. Only the applied static force, as shown in Figure 4.2, will be considered.

2. *Exact geometric description* The dimensions of the block given in Figure 4.2 will be accepted as shown. A half-symmetric model will be formed from the block. A quarter-symmetric model could be formed but it would place the location of anticipated maximum stress at the base of the notch, along a cut boundary. Because the geometry of this example is more complex than the plate

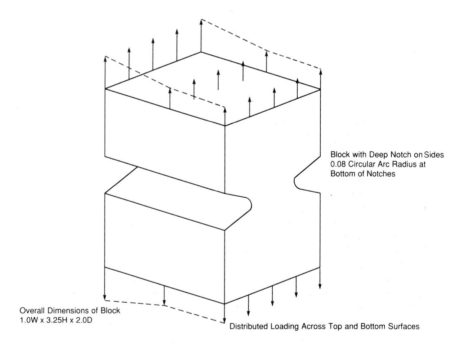

Block with Deep Notch on Sides
0.08 Circular Arc Radius at
Bottom of Notches

Overall Dimensions of Block
1.0W x 3.25H x 2.0D

Distributed Loading Across Top and Bottom Surfaces

Figure 4.2 Notched block: Problem definition.

with the circular hole, it was decided not to use a quarter-symmetric model. The force profiles are symmetric about the center
plane so that the force is not a consideration in the decision as to
what type of symmetry may be used.

3.. *Forces* The forces shown in Figure 4.2 are uniaxial tensile
forces. Two cases are shown. The first is a uniform distribution
and the second is a nonuniform distribution, symmetric about the
center plane. For the first case, a two-dimensional model would
be adequate. Because the block is fairly deep, a plane-strain
model will be used. In the case of the nonuniform force distribution, a two-dimensional model would give an approximation of the
true stress. However, a full three-dimensional model will also be
used.

4. *Constraints* Roller constraints will be used along the center plane of symmetry. To prevent numerical problems, constraints will be applied to the bottom surface of the block, in lieu
of an equal and opposite set of applied forces. Along the bottom
face, roller constraints will be applied in the vertical direction.
The surface will be free to move in the two, in-plane directions.
One point on the block will have to be constrained, normal to the
plane of symmetry to prevent rigid body motion of the block.

4.2.3 Example: Pressure Vessel

A simple pressure vessel is shown in Figure 4.3. The pressure
vessel has an internal pressure of 100.0. The pressure vessel
can be considered to be a solid of revolution about its axis and
symmetric about a center plane, normal to its axis.

1. *Objectives* To solve for the maximum static stress in the
pressure vessel due to the internal pressure force. In this example, thermal effects will be ignored. This example will be modified in Chapter 12 to illustrate the inclusion of thermal effects.

2. *Exact geometric description* The dimension as shown in
Figure 4.3 will be used as shown. In a real pressure vessel, the
locations of welds should be noted to see if they could cause a
geometric discontinuity in the structure. Lifting lugs and ex-
terior attachment points would also have to be noted. This pressure vessel will be modeled as an axisymmetric structure. The
center plane of symmetry will be used to model only half of the
pressure vessel.

3. *Forces* Internal pressure will be the only applied force.
Other forces that could also be considered under certain circum-
stances are gravity force and reaction forces at support points.
Fluid dynamic forces, either gas or liquid, could also be taken

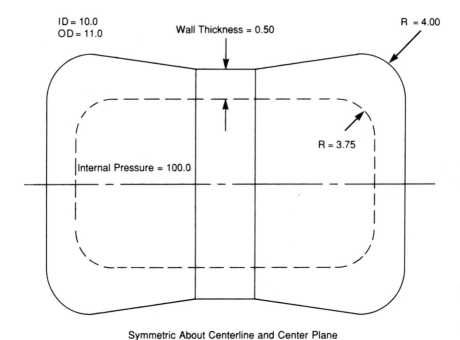

ID = 10.0
OD = 11.0

Wall Thickness = 0.50

R = 4.00

R = 3.75

Internal Pressure = 100.0

Symmetric About Centerline and Center Plane

Figure 4.3 Pressure vessel: Problem definition.

into account in the event of any agitation or reaction of the vessel's contents. These internal forces, if they are applicable, are more difficult to quantify and require a detailed understanding of the process taking place within the vessel.

4. *Constraints* The only constraints that need to be applied are roller constraints in the axial direction along the cut center plane of symmetry. Because this is an axisymmetric model, constraints in the radial direction are not required so long as the boundary at the centerline has a radial coordinate of 0.

4.2.4 Example: Crank

The hand crank is a welded machinery component as shown in Figure 4.4. A handle is connected to one end of the crank through a 0.625 reamed hole and a shaft is connected to the other end, and torque is transmitted by a square key. Hubs at each end of the crank are welded to a baseplate.

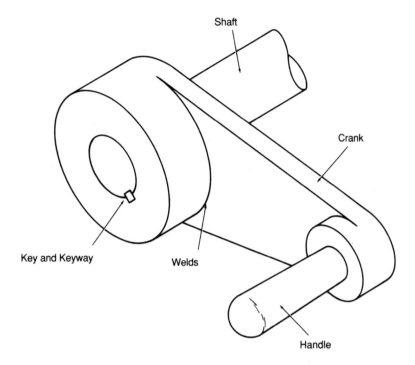

Figure 4.4 Crank: Problem definition.

1. *Objectives* The objective of the analysis is to determine the maximum stress at the keyway (the point of maximum anticipated stress) as a function of transmitted torque. Once again, thermal and dynamic effects will be neglected. It will be arbitrarily decided that the only part of the assembly to be considered will be the crank. Stresses in the shaft will not be calculated in this example.

2. *Exact geometric description* From an examination of the detailed part drawings, it is determined that the handle is attached to the crank with a light interference fit (0.0002—0.0010 tight). From this, it may be assumed that there is complete contact between the handle and crank. At the shaft end, the fit between the shaft and crank bore is from 0.0005—0.0010 loose. Therefore, in a worse-case situation one would have to assume that there is no torque transmitted due to friction between the crank and shaft and that all the torque is transmitted by the

key. The key is a tight fit in the shaft and a loose fit (0.0002–0.0008 loose) in the crank. Because the crank only transmits force in one direction of rotation, it may be assumed that there is complete contact between the driving side of the key and the corresponding side of the keyway. It may be further assumed that because the crank does not alternate directions, there will be no impact-type force between the key and keyway.

3. *Forces* Applied forces will consist of the torque force applied as a linear distributed force at the handle.

4. *Constraints* Displacement constraints are applied at the shaft/keyway area of the crank. Local radial constraints are applied around the inside of the bore of the crank to prevent deformation, but not to transmit any torque. Local tangential constraints are applied to one edge of the keyway to obtain appropriate reaction forces.

Although this could be approximated as the two-dimensional problem, a full three-dimensional model will be formed. It is not intuitively obvious whether or not there will be a force gradient along the length of the keyway. The highest force may be transmitted at the base of the crank where the plate attaches to the hub. However, this area has the most reinforcement due to the plate so that the highest stresses may not necessarily occur there. For that reason, a full three-dimensional model of the crank will be used. The handle, shaft, and key will not be included in the model.

4.2.5 Example: Gear Segment

This example problem may be approached from a number of different directions with various objectives and levels of accuracy desired. The gears to be modeled are shown in Figure 4.5. They are a pair of straight spur gears with a 20° pressure angle and a dimetral pitch of 2. Both the diving and driven gear have 40 teeth each and, therefore, operate on a pitch diameter of 20.0. All other dimensions are standard for full-depth teeth of this pitch and pressure angle.

Various options for modeling from the simplest to the most complex might be the following:

Two-dimensional model of a single tooth with applied forces.

Two-dimensional model of a segment of teeth with applied forces. This would allow for the interaction of the forces on adjacent teeth to be more correctly accounted for in the fillet area between the teeth.

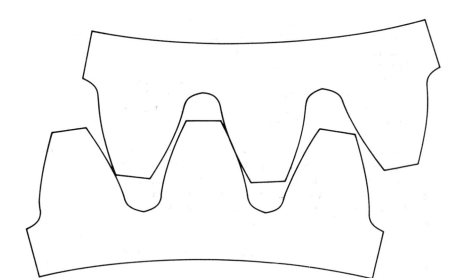

Straight Spur Gears

Pitch Dia. = 20.0

Pressure Angle = 20 degrees

Number of Teeth (Each Gear) = 40

Figure 4.5 Two gears in contact: Problem definition.

Two-dimensional model of two segments, one of the driving gear and the other of the mating, driven gear, with gap elements between the contacting profile surfaces to allow the force between the teeth to be more correctly distributed by the model.

Three-dimensional model of a single tooth with applied contact pressure distribution to be applied. This would allow for a variation in contact pressure and stress along the length of the tooth.

Three-dimensional model of a segment of teeth.

Three-dimensional model of contacting segments of driving and driven gears with gap elements in between mating tooth surfaces. This would allow for lengthwise variation in contact due to "crowning" of the teeth. Because the force between the teeth is constantly changing as the teeth roll through contact, a segment model of this type should be incremented to simulate a stress time history.

Three-dimensional model of driving and driven segments with allowance for mounting errors and mounting flexibility. There is normally some flexibility associated with the gear mountings and bearings that can cause the contact pattern to shift and thereby change the pressure and stress distribution.

The purpose of listing the above options is not to confuse the matter, but to illustrate some of the thinking that must take place long before the actual finite element modeling begins. In this gearing problem, there are more approaches to the problem than in the previous cases. There is obviously no one "cookbook" method of modeling a gearing problem.

All of the above models are for static deflection and stress analysis. Thermal effects are generally not considered to be significant in gearing problems except in unusual circumstances. Dynamics are taken into account in the event of a resonance condition of the gear drive system, either lateral or torsional, excited by a harmonic of the running speed of one of the gears or by the gear set's tooth mesh frequency.

1. *Objectives* There may be one of several objectives to the analysis, each of which may require a different type of model. The two most common objectives of gearing analysis are the determination of root fillet bending stresses that can lead to fatigue or the determination of profile contact stresses that can lead to spalling or pitting. For the present purposes, the objective will be to obtain the root fillet cyclic bending stresses.

2. *Exact geometric description* All dimensions such as addendum, dedendum, etc. will be for standard 20° pressure angle, 2 pitch spur gears with 40 teeth. No lengthwise crowning or profile modifications will be assumed. The teeth will be assumed to have a perfect involute shape with circular arc fillets and no undercut in the fillet.

Stress analysis of gear teeth can be quite complex. A number of simplifying assumptions are made here. However, this type of problem description and model should be quite adequate to represent first-order effects and give a result equal to or better than conventional closed-form solutions.

4.2.6 Example: Turbine Blade

This example case is a steam turbine blade from a low-pressure stage of a large power-generation turbine as shown in Figure 4.6.

A typical analysis for a blade of this type would include a static stress calculation, natural frequency and mode shape calculation, and possibly a calculation of dynamic stresses. Most of the static stress in the blade is from centrifugal force, due to the blade rotating at 3600 rpm. Steam forces are typically two orders of magnitude less than the centrifugal forces. Steam forces may be in the 100–300 lbf range, whereas the centrifugal force may be 10,000–30,000 lbf at the base of the blade.

There are several parts to a turbine blade. The airfoil section of the blade is tapered and twisted with highly curved surfaces. The turbine blade is sensitive to the interaction of its natural frequencies with harmonics of the running speed. Therefore, it is important to correctly calculate the natural frequencies, which, in turn, requires an accurate geometric model of the airfoil section. In some cases, the blades are independent of each other, i.e., freestanding. In other designs, the blades are connected at their tips into groups of three to seven, typically. The blades are grouped for dynamic stiffening to minimize dynamic response and also to provide a sealing surface to prevent steam leakage. The cover is attached to the blade group by a cold rivet, known as a tenon that is integral with the airfoil. As may be expected, the junction of the tenon with the top of the airfoil is an area of highly localized stress, especially when the residual riveting stresses are included. The blade is attached to the rotor by a slotted, "firtree"-type construction so that each blade fits into a precisely machined slot in the rotor. The attachment region of the blade has high, localized, static stresses in the "hook" notches. In these areas, specification of the appropriate contact between the blade hooks and rotor serrations is important to the calculation of local stress that may approach or equal the yield strength of the material.

1. *Objectives* The objective of this study is to calculate the maximum static stress in the blade under steam loading and at operating speed. Any consideration of dynamic effects will be deferred to Chapter 11. Thermal effects will not be considered here. This low-pressure blade operates at 150°F nominally with changes of ±20°F.

2. *Exact geometric description* Dimensions are critical in two areas: The first is the blade/rotor attachment areas where there are six contact surfaces. Due to a buildup of tolerances, there may occur conditions when all six surfaces are not in contact and only three or four surfaces are carrying the load. If the surfaces are slightly out of contact, i.e., within 0.001 in., then elastic deformation will pull the out-of-contact surfaces into contact. Even so, the notches associated with the surfaces originally in contact

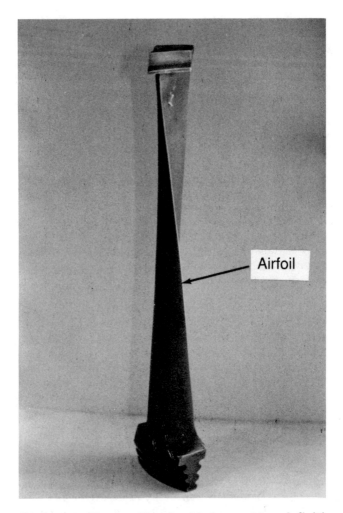

Figure 4.6 Above: Turbine blade; problem definition. Opposite, top: Turbine blade; showing cover area. Opposite, bottom: Turbine blade, showing root attachment area.

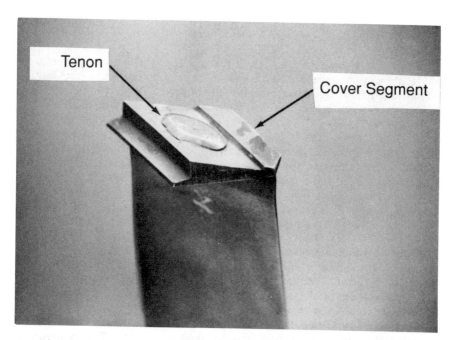

Tenon

Cover Segment

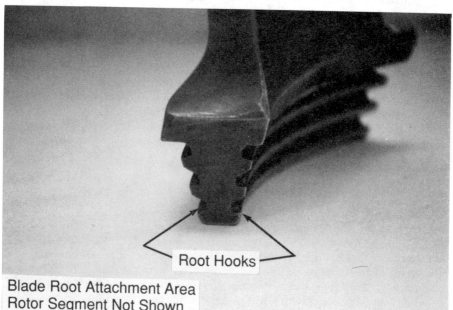

Root Hooks

Blade Root Attachment Area
Rotor Segment Not Shown

will see a higher stress than notches associated with surfaces initially out of contact. In this area, the exact geometry may be obtained from the detail part drawings that include the profile tolerances or from measurements of the actual blade and rotor slot.

The second area is the airfoil profile. This is important in the calculation of natural frequencies and dynamic response, but not as important in the calculation of static stresses. Again, the profile description, generally in terms of coordinates at a number of cross sections, may be obtained from the manufacturing drawings or from the actual blade.

3. *Forces* Two types of forces are applied to the turbine blade, the centrifugal force and steam forces. Centrifugal forces will be calculated automatically by the finite-element program so that the only input parameter that needs to be defined is the rotational speed. Dimensions of the blade must be defined in terms of their absolute distance from the rotor centerline, the center of rotation. So long as the geometry input and material density are properly specified, the centrifugal body forces will be correctly calculated and applied in the finite-element model. The applied steam forces may be applied either as a distributed pressure force across the airfoil or, in this case, resolved into tangential and axial force components based on force distributions in the two directions and distributed across the appropriate airfoil nodes.

4. *Constraints* Because the areas of anticipated highest stress are in the firtree attachment, a section of the rotor surrounding the blade attachment is included in the model. Displacement constraints are applied at the outside boundary of the rotor segment. The contact surfaces of the blade and rotor are coupled.

This is an obvious case for a full three-dimensional model. The use of a two-dimensional refined model in the attachment area will be discussed in Chapter 10.

4.3 FREE-BODY DIAGRAM

As an intermediate step following the problem definition and before actual model coding, it is advisable to construct a free-body diagram of the portion of the structure that is to be included in the model. This free-body diagram should show applied forces and displacement constraints in the same location, direction, and magnitude as they will be applied to the finite-element model. The construction of the free-body diagram allows for a check on those items such as forces and constraints that need to be quantified prior to model coding. Because forces and displacements can only be applied at nodes, it is necessary to have the locations of all forces and displacements identified prior to model generation to insure that nodes are positioned correctly. With modern CAD and solid modeling software, it is possible to create a free-body diagram within the computer. Otherwise, it is necessary to create the free-body diagram on paper that can then be copied as many times as necessary for the next step, gridwork development, and coding.

4.4 FREE-BODY DIAGRAM: EXAMPLE CASES

4.4.1 Example: Plate with Circular Hole

A free-body diagram for the plate with a circular hole is shown in Figure 4.7. The diagram shows the quarter-symmetric portion of the plate. The diagram is shown as a two-dimensional representation of the plate because there are no out-of-plane forces and should not be any out-of-plane displacements. Roller constraints are specified along both the cut boundaries. Along

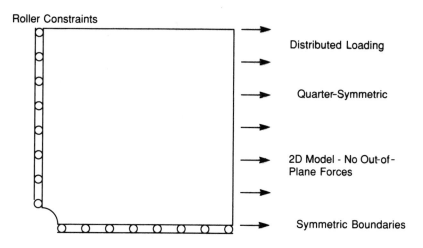

Figure 4.7 Plate with circular hole: Free-body diagram.

the X axis, it is necessary to allow motion in the X direction because of the applied force in that direction. Along the Y axis,
there should be no displacement in the X direction because of
the balanced forces and assumption of symmetry. There will be
displacement in the Y direction along the Y axis due to Poisson's
effect as the plate becomes narrower due to the tension in the X
direction. The applied force is shown as a uniform force distributed along the right-hand edge of the plate acting only in the X
direction.

4.4.2 Example: Notched Block

The free-body diagram of the notched block is shown in Figure
4.8. The diagram shows a half-symmetric representation of the

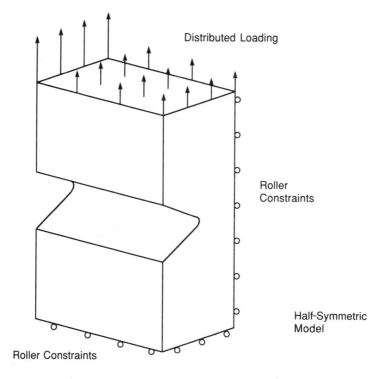

Figure 4.8 Notched block: Free-body diagram.

block with roller constraints along the plane of symmetry constraining against out-of-plane deformation. Roller constraints are also shown on the bottom surface of the block opposite the end of the applied force. These roller constraints arise in lieu of applying an equal and opposite force distribution.

A distributed force is applied across the top of the block. Because the force distribution is constant in the Y direction, the use of symmetry is acceptable.

4.4.3 Example: Pressure Vessel

The free-body diagram for this axisymmetric model is shown in Figure 4.9. Because of the symmetry about the midplane, only half of the pressure vessel is included in the model. Roller constraints to prevent against deformation in the axial direction are used on the cut plane. No constraints are required for the radial direction for an axisymmetric model. A uniform internal pressure is applied to the entire inside surface of the pressure vessel.

4.4.4 Example: Crank

In modeling the crank, it was decided to only include the crank itself in the model and not the handle, shaft, or key. The free-

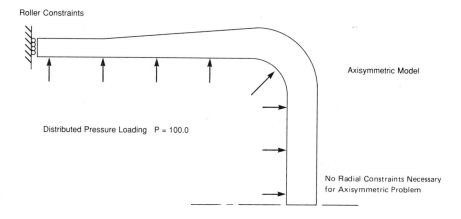

Figure 4.9 Pressure vessel: Free-body diagram.

body diagram of the crank is shown in Figure 4.10. Displacement constraints are applied at the inside of the shaft bore to represent the reaction of the shaft and key. The effect of the shaft is represented by allowing displacement in the local tangential direction on the inside of the shaft bore, but constraining against any displacement in the local radial direction. The key is represented by displacement constraints on the side of the keyway in the tangential direction. The handle is represented by applied forces at the inside of the handle bore. Symmetry cannot be used here to reduce the model size due to the axisymmetric location of the keyway.

An alternative free-body diagram would include the shaft and key. This would require, however, the use of gaps between the separate components, which causes the problem to become nonlinear and substantially increases the computer runtime.

4.4.5 Example: Gear Segment

The gear segment is shown as two free-body diagrams. The two diagrams are shown in Figure 4.11 for a single tooth and in Figure 4.12 for a three-tooth, driven-gear segment. If we assume

Only the Crank Is Included in the Free-Body Diagram

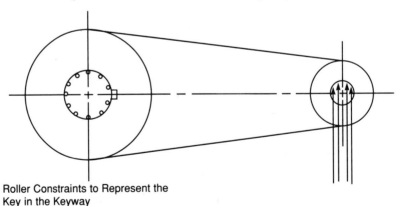

Roller Constraints to Represent the
Key in the Keyway

Distributed Loading to
Represent the Handle

Figure 4.10 Crank: Free-body diagram.

Distributed Loading Along Pressure
Side of Tooth Profile

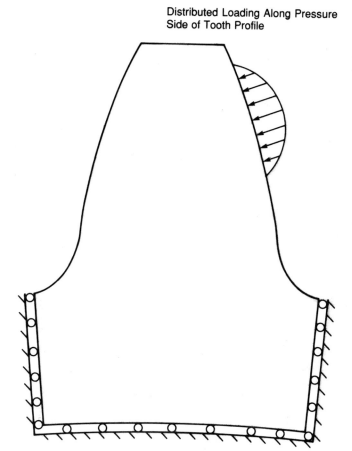

Roller Constraints Along Cut Boundaries

Figure 4.11 Single gear tooth: Free-body diagram.

that there is no lengthwise curvature (crowning) of the teeth,
then the free-body diagrams apply to either the two- or three-
dimensional cases.

The free-body diagram of the single tooth shown in Figure
4.11 shows a force distribution on the pressure side of the tooth
profile. Displacement roller constraints are applied at the base

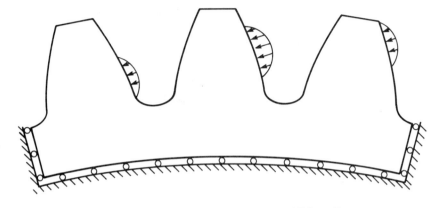

Roller Constraints Along Cut Boundaries of Driven Gear

Figure 4.12 Three-tooth gear segment: Free-body diagram.

of the tooth in the radial direction. Roller constraints are also applied to the sides of the tooth along the cut boundaries between the teeth in the tangential direction.

The free-body diagram for the three-tooth segment shows distributed forces across the pressure faces of the three teeth. Displacement constraints are applied to the cut boundaries of the driven gear. Roller constraints are again applied at the bottom of the segment in the radial direction. Roller constraints in the tangential direction are applied at the cut boundaries of the first and third driven gear teeth in the tangential direction.

4.4.6 Example: Turbine Blade

The free-body diagram for a grouped turbine blade is shown in Figure 4.13. The free-body diagram consists of the blade airfoil, integral tenon, a portion of the cover, root attachment, and a segment of the mating rotor including the rotor slot into which the blade fits. The portion of the cover was included in the free-body diagram in order to provide the proper centrifugal force to the tenon and so as not to require specifying boundary conditions directly at the highly stressed tenon. The rotor segment is included in the model for a similar reason: To eliminate the need to apply assumed boundary conditions at the load-bearing surfaces of the root hooks.

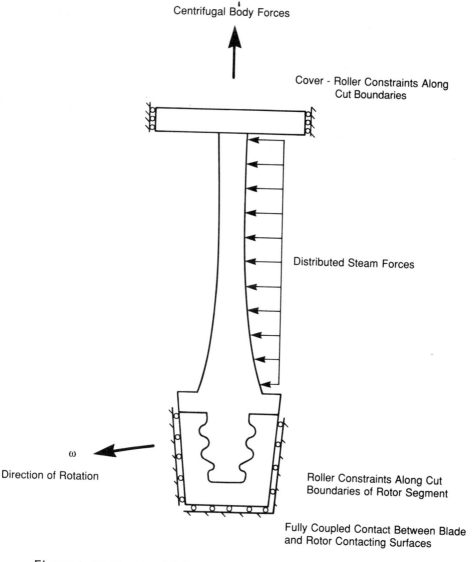

Figure 4.13 Turbine blade: Free-body diagram.

Applied forces consist of a distributed force in the tangential and axial directions along the airfoil section representing the applied steam forces. Displacement constraints are applied at the cut boundaries of the rotor segment with roller constraints in the radial direction and along the side of the rotor segment in the tangential direction. The contact surfaces between the blade root and rotor are assumed to be coupled.

4.5 SUMMARY

The process of forming a finite element model can be made more straightforward by clearly defining the problem prior to sitting down at a computer terminal and entering input data. This chapter covers steps 1 and 2 of a three-step procedure for developing a model. Step 1 involves defining the engineering problem and objectives of the analysis. Step 2 involves a decision on how much of the structure to include in the model and how much can be cost-effectively included in the model. Step 2 also covers the specification of boundary conditions. Once these two steps have been completed, the actual development of the finite element model may proceed in an organized manner. Time spent in defining and specifying the analysis will pay dividends later on in the actual finite element calculations and data analysis.

REFERENCE

1. Seeley, F. B. and J. O. Smith, *Advanced Mechanics of Materials*, 2nd ed., Wiley, New York, 1932.

5

FINITE ELEMENT MODEL

Once the problem has been defined and the free-body diagram
constructed, the development of the finite element gridwork,
coding, and entering the model into the computer consist of (1)
dividing the free-body diagram into elements, (2) applying forces
to nodes, and (3) applying displacement constraints to nodes. Of
the three tasks, division of the free-body diagram into elements
is by far the most difficult and at the heart of the modeling process. The specification of applied forces and displacement constraints should be virtually automatic if the free-body diagram is
correctly prepared.

This process of actually creating the finite element gridwork
with boundary conditions and entering the data into the computer
used to be a multistep procedure. First, the gridwork was laid
out on paper; next, lists of the nodes, elements, and boundary
conditions were prepared; the input data were either keypunched
onto punch cards or entered into a data file; and check runs were
made to look for coding and data input errors. With the use of

computer graphics, interactive model generation, and solid modeling software, the process is streamlined. The overall gridwork may still be sketched out on paper; however, absolute accuracy is not required. Key nodes need to be identified and their coordinates listed. Data input and error checking are performed simultaneously. With the use of interactive graphics, input data can be plotted back and corrected as it is entered, thereby reducing the total time for a completely debugged model to be available for analysis. When a computer-aided design (CAD) system is used to provide geometric data to the finite element model, the step of extracting key node data and entering it into the computer is eliminated. The key points, lines, arcs, areas, and volumes can be manipulated by either the CAD or finite element modeling software prior to actually defining the gridwork.

Whether the model is generated by hand or with the assistance of modeling software, the same basic principles apply to achieve the most cost-effective model. Some general guidelines for the breakup of the model into elements are given in Table 5.1. These guidelines are expanded in the subsequent sections.

5.1 ELEMENT DENSITY

In dividing the gridwork (free-body diagram at this stage) into elements, two concepts are important: absolute element density and relative element density. *Absolute element density* refers to the total number of elements in the gridwork. *Relative element density* refers to the distribution and size of elements in various parts of the gridwork. Absolute element density is generally a tradeoff between accuracy and computer use.

Several parameters are affected by either absolute or relative element density or both

Absolute accuracy.
Computer runtime and storage.
Element distortion.
Geometry definition.
Stress gradient definition.

All of the above items except computer runtime and storage are related to accuracy.

For a given model and set of loading conditions, the accuracy may be expressed as a function of three main parameters: element type, number of elements, and element distortion. A general expression for accuracy can be given as

Table 5.1 Guidelines for the Division of the Model into Elements

Element types must be consistent, i.e., three- and two-dimensional elements cannot be mixed in the same analysis. Plate and beam elements can be mixed with three-dimensional solid elements when the rotational degrees of freedom are accounted for.

More and smaller elements should be used in the areas of anticipated high-stress gradients such as notches, fillets, and holes.

Extremely fine mesh should be used when forces must be applied near high-stress areas. Applied forces tend to distort the stress distribution near the point of application.

The ideal goal in developing the gridwork is to have as uniform as possible a change in stress between adjacent elements throughout the model. In other words, the relative element density should follow the stress distribution.

Quadrilateral elements should be as square as possible and triangular elements should be equilateral whenever possible, especially in critical areas.

Gross element distortion should be avoided anywhere in the model, even in noncritical areas.

Triangular and wedge elements should be used in transition areas between coarse and fine meshes.

Adjoining elements must share common nodes and common degrees of freedom.

$$A \; \alpha \; \frac{T \; N^{1/n}}{D}$$

where

A = accuracy, where 1.00 represents perfect accuracy and 0 represents infinite error

N = the number of elements in the model or a specific region of the model

T, n = parameters reflecting the integration order of the element (linear, quadratic, etc.)

D = a parameter reflecting the degree of distortion in the model or a specific region of the model

The parameters T and n adjust for the fact that higher-order elements, such as quadratic elements with midside nodes, may achieve the same accuracy using fewer elements as linear strain elements with more elements. All models, regardless of what kind of elements are used, should asymptotically approach almost perfect accuracy as the number of elements approaches infinity. Therefore, the effect of higher-order elements may be more pronounced for models in which there are fewer elements. Properties of various common element types and rates of convergence are discussed more thoroughly in Chapter 7. As an example, however, the convergence of two triangular, two-dimensional elements is shown in Figure 5.1. The model is a simple cantilever beam in bending. The two elements are a three-node linear displacement, constant strain triangle and a six-node quadratic displacement, linear strain triangle. It can be seen that for a few elements, there is a significant difference in the accuracy of the two elements, but both elements converge to about 0.98–0.99 accuracy with more than 50 elements. It must be noted that this simple comparison does not account for computer runtime in which the higher-order more powerful elements require more computer use per element.

The accuracy of the model increases with the number of elements for several reasons.

1. The structure's geometry can be more accurately defined with more points. This is especially true when a curved boundary is modeled with linear elements having nodes only at their vertices. These elements must represent the curved boundary as a series of straight lines.

2. Stress gradients can be more accurately defined by having more elements, thereby minimizing stress gradients within elements. This is especially important for linear, constant strain elements.

3. More elements generally means smaller elements close to a fillet or notch surface where a maximum stress may occur. Smaller elements means that the centroids of the surface elements will be closer to the structure's surface simply because of the element size. More elements in these regions give more definition of stress gradients.

When a structure has a simple shape and simple loading, the number of elements may not have any real effect of accuracy. In the simplest cases, almost perfect accuracy can be achieved with one element. Of course, the model has to match exactly the assumptions and type of analysis built into the element. For example, one beam element should give virtually perfect correlation

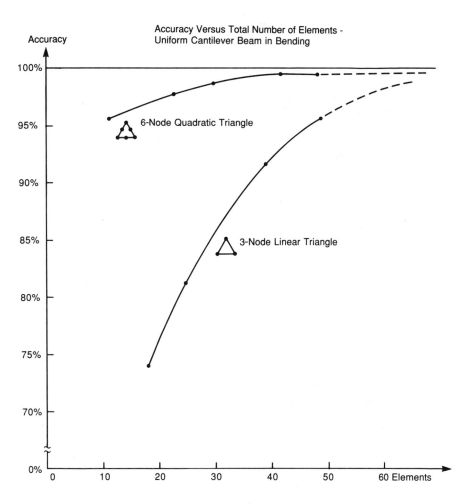

Figure 5.1 Convergence of two triangular element types.

when compared to a closed-form solution from conventional beam theory. In the same way, a two- or three-dimensional solid element in simple tension with the appropriate boundary conditions should also give perfect results. Due to computer truncation error, it is never possible to have "perfect" agreement. However, perfect or "almost perfect" accuracy can be interpreted as being agreement to five or six significant figures.

5.2 ELEMENT DISTORTION

Element distortion is given in this chapter as a single parameter;
however, its effect on accuracy is complex. The quantitative
effects of distortion depend on the type of element, geometry
of the structure, type of loading, and specific type of distortion. Chapter 8 is devoted entirely to the topic of element distortion.

Element distortion can often be reduced simply by using more
elements in part of a model, i.e., higher relative element density. Quite often, unacceptable element distortion is the result
of trying to have a fine gridwork in one part of a model and
reducing too quickly to a coarse gridwork in the noncritical
areas. This can be resolved by adding more elements in the
noncritical areas, thereby reducing the difference in the relative element density.

Examples of gross element distortion that must always be
avoided are illustrated in Figure 5.2. The types of distortion
shown would cause a fatal error in virtually any finite-element
program.

5.3 BOUNDARY CONDITIONS

The specification of displacement boundary conditions is straightforward if they have been specified on the free-body diagram.
In generating the gridwork, node locations may have to be adjusted and additional nodes and elements added to have nodes
available for constraints to be applied at the proper location.
Nodes must be positioned at the locations along the boundary that
correspond to constraint locations in the actual structure. It is
never acceptable to move a constraint to match a node; rather,
the node must be located to match the actual constraint.

The specification of applied forces follows the same basic rules
as constraints with respect to node location. Forces should be
transferred from the free-body diagram to the appropriate nodes
and degrees of freedom. Where there are distributed pressure-
type forces, the distribution may be input directly, if the program will allow it, or the equivalent nodal forces must be calculated according to the rules given in Chapter 3. As mentioned above, when stresses near a location of applied force are
of interest, it may be necessary to use several layers of small
elements in the area to get a reasonable resolution of the localized stresses.

Elements Should Never
Double back on Themselves

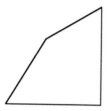

An Element Corner Should
Never Collapse to 180 Degrees

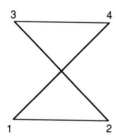

Element Numbering Must Be
Either Clockwise or
Counterclockwise

Figure 5.2 Examples of gross element distortion.

5.4 FINITE ELEMENT MODEL: EXAMPLE CASES

For the following examples, data on computer runtimes for the
various models are presented. These runtimes are dependent on
the computer hardware and finite element software being used.
Therefore, the data are presented for a relative comparison between the various models within an example case and between example cases.

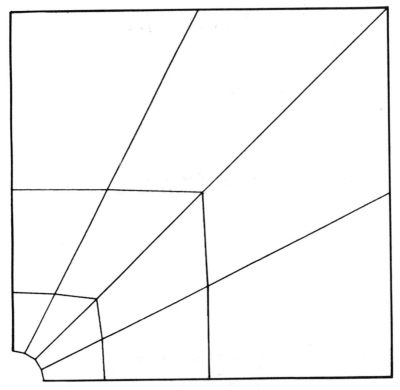

Coarse Gridwork 12 Elements
No Attempt Made to Optimize Gridwork

Figure 5.3 Plate with circular hole: Case 1 gridwork.

5.4.1 Example: Plate with Circular Hole

Figure 5.3 shows the first, coarse gridwork generated for the
plate with circular hole. Because this is a thin plate and has
been determined to be a two-dimensional problem, two-dimensional plane stress elements are used.

It is anticipated that the highest stresses and highest stress
gradient will occur at the edge of the hole at the $X = 0$ location.
The smallest elements are located around the boundary of the
hole. These nodes and elements can be easily generated by defining a local, cylindrical coordinate system with its origin at the

center of the hole. Along the Y axis, close to the point of highest anticipated stress, it is quite easy to have uniform elements.
A total of 12 elements are used in this coarse grid model.

The stress and stress gradient can be anticipated by using a
closed-form solution [1] to the problem; this gives the stress in
the vicinity of the hole as

$$\sigma = \frac{\sigma_0}{2} \left\{ 2 + \frac{\rho^2}{x^2} + 3\frac{\rho^4}{x^4} \right\} \tag{5.1}$$

where

ρ = the radius of the hole
x = the distance from the center of the hole

The maximum stress is then $3\sigma_0$.

Forces are applied to the nodes along the right-hand end in
the manner prescribed by the free-body diagram described in
Chapter 4. The corner node receives half of the force of the
other nodes. Note that the node at the X axis receives the same
force as the others because it is assumed that the distributed
force is continuous along the right-hand edge.

Constraints are specified at the nodes along the cut boundaries. Along the X axis, nodes are constrained in the Y direction only, and along the Y axis, nodes are constrained along the
X axis only.

The result of the first calculation compared to the closed-form
maximum stress is

Finite element maximum centroidal normal stress = 130 psi
Finite element maximum nodal stress   = 191 psi
Closed-form calculated maximum normal stress   = 300 psi

To test the sensitivity of the calculation to element density, a
second and third calculation are performed with the gridworks
shown in Figures 5.4 and 5.5. For the second case (fine mesh),
132 elements were used. In the third case (selective mesh), 84
elements are used. In the third case, the total number of elements is less than for the second case; however, the relative
(local) element density around the hole is about the same. Forces
and constraints are specified in a manner similar to that of the
first calculation.

The results of the calculations performed with the three models are shown in Table 5.2, along with a comparison of accuracy
and computer runtime. These results can also be shown graphi-
cally in Figures 5.6—5.8. These results show that although the

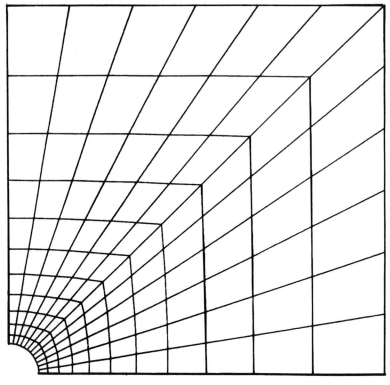

Fine Gridwork   132 Elements
High Element Density Throughout

Figure 5.4 Plate with circular hole: Case 2 gridwork.

second case had the highest number of elements, the third case
had equal accuracy when the elements were concentrated around
the hole. In this case, the accuracy is defined as the ratio of
maximum stress around the hole to the closed-form calculated
maximum, 300 psi.

5.4.2 Example: Notched Block

A total of three separate finite-element models are generated for
this example. In section 4.4.2, the two free-body diagrams,
two-dimensional and three-dimensional, are discussed. One two-

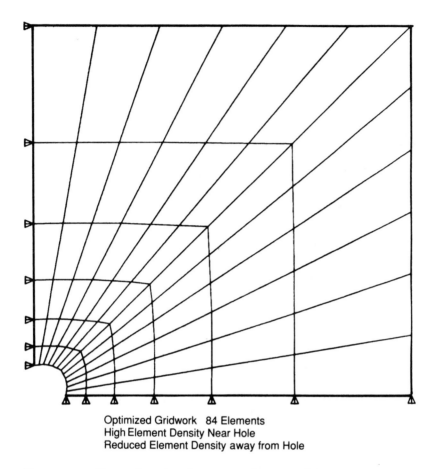

Optimized Gridwork 84 Elements
High Element Density Near Hole
Reduced Element Density away from Hole

Figure 5.5 Plate with circular hole: Case 3 gridwork.

dimensional finite element model is constructed and shown in Figure 5.9, and two three-dimensional finite-element models are constructed and shown in Figures 5.10 and 5.11. The two-dimensional model contained 38 plane strain elements. The coarse three-dimensional model contained one layer of 24 three-dimensional solid elements and the fine three-dimensional model contained two layers of 38 elements (the same pattern as in the two-dimensional model) for a total of 76 elements. The details of the three models and their computer runtimes are given in Table 5.3.

Table 5.2 Results of Plate Calculations

	Case 1 coarse	Case 2 fine	Case 3 intermediate
Centroid location from edge of hole	0.462	0.0697	0.0674
Normal stress at centroid	130.61	231.79	229.63
Exact solution stress at centroid location	124.45	227.50	229.30
Nodal stress at edge of hole	191.36	300.60	296.36
Exact solution stress at edge of hole	300.0	300.0	300.0
Number of elements	12	132	84
CPU time, sec	11.36	76.74	49.96

There are only two cross sections of finite element mesh used for the three models. The two- and three-dimensional fine models use the same cross-sectional mesh pattern. In both patterns, the highest element density is concentrated around the bottom of the notch where it is anticipated that the largest stresses and highest stress gradients are located. The mesh layout is begun around the notch and elements maintained as square as possible. The remainder of the gridwork is developed to fill in the balance of the cross section. Triangular cross-section elements are used to illustrate the concept of a transition between a fine and coarse mesh. If we assume that the maximum stresses occur at the notch fillet and that the determination of maximum stress is the objective of the analysis, a minimum number of elements is used in the balance of the block.

A comparison of the two- and three-dimensional fine models shows that for the same cross-sectional element density, almost 15 times as much computer runtime is required (248 sec vs. 16.6 sec) for the three-dimensional model as for the two-dimensional model. This shows the motivation behind using a two-dimensional

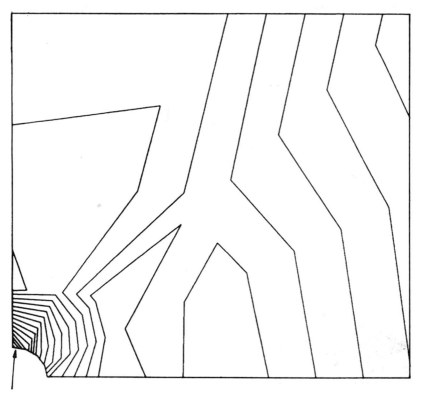

Location of Maximum Principal Stress
Maximum Nodal Stress = 191.36
Closed-Form Solution = 300.00

Coarse Gridwork Has Too Few Elements to
Give Accurate Stress Results at Hole

12 Elements => Run time = 11.36 sec

Figure 5.6 Plate results: Case 1.

approximation of a three-dimensional loading distribution. The
three-dimensional coarse model requires only about one-third of
the computer runtime of the three-dimensional fine model; however, with only two planes of nodes on the top surface, it is impossible to provide a really accurate representation of the loading distribution.

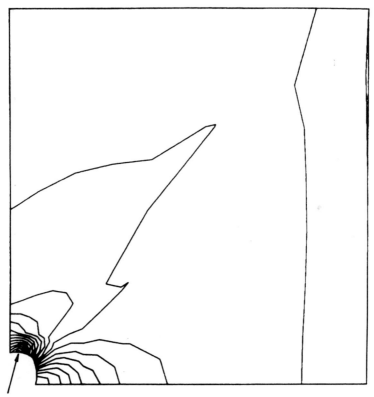

Location of Maximum Principal Stress
Maximum Nodal Stress = 300.60
Closed-Form Solution = 300.00

132 Elements => Run time = 76.74 sec

Fine Gridwork Gives Excellent Agreement with Closed-Form
Solution, But with 7 Times the Computer Time

Figure 5.7 Plate results: Case 2.

Roller constraints are applied to the nodes on the boundaries
as discussed in Section 4.4.2 and shown on the free-body diagrams, Figures 4.8 and 4.9. For the two three-dimensional cases,
one node is constrained in the Z direction to prevent rigid body
motion. The two sets of roller constraints should prevent all

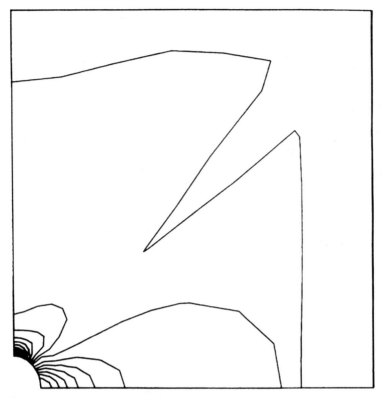

Location of Maximum Principal Stress
Maximum Nodal Stress = 296.36
Closed-Form Solution = 300.00

84 Elements => Run time = 49.96 sec

Optimized Gridwork Gives Good Agreement with
Significantly Less Computer Time than Fine Gridwork

Figure 5.8 Plate results: Case 3.

other translations and rotations so that the only remaining dis-
placement to be accounted for is rigid-body displacement in the
Z direction. The selection of the node to be constrained in the
Z direction is entirely arbitrary.

Forces are applied to the nodes on the top of each block model
according to the distribution shown in the free-body diagrams,

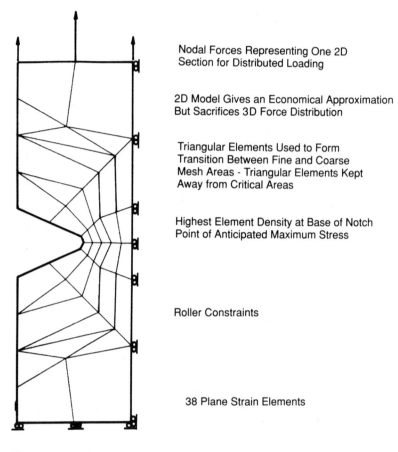

Nodal Forces Representing One 2D
Section for Distributed Loading

2D Model Gives an Economical Approximation
But Sacrifices 3D Force Distribution

Triangular Elements Used to Form
Transition Between Fine and Coarse
Mesh Areas - Triangular Elements Kept
Away from Critical Areas

Highest Element Density at Base of Notch
Point of Anticipated Maximum Stress

Roller Constraints

38 Plane Strain Elements

Figure 5.9 Notched block: Two-dimensional model.

Figures 4.8 and 4.9. Forces are assigned to each node according to the integration of the distributed force over the area of
each surface node. The force distributions for both the two-
and three-dimensional models are shown in Figure 5.12.

Stress results are provided in Table 5.4. These results are
presented in terms of the maximum centroidal element stress and
maximum nodal stress for each case. As expected, the direction
of the maximum stress is in the Y direction.

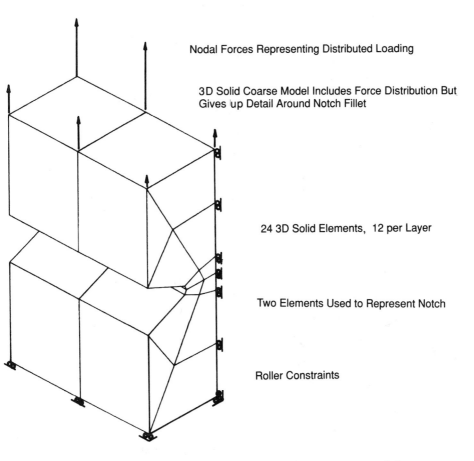

Nodal Forces Representing Distributed Loading

3D Solid Coarse Model Includes Force Distribution But Gives up Detail Around Notch Fillet

24 3D Solid Elements, 12 per Layer

Two Elements Used to Represent Notch

Roller Constraints

Figure 5.10 Notched block: Three-dimensional coarse model.

Stress contour plots are shown in Figure 5.13 for the two-dimensional model, in Figure 5.14 for the coarse three-dimensional model, and in Figure 5.15 for the fine three-dimensional model. These three figures show that the same basic stress distribution exists for all three models. The difference is the degree of resolution leading to the determination of maximum stress and the refinement of the stress gradients.

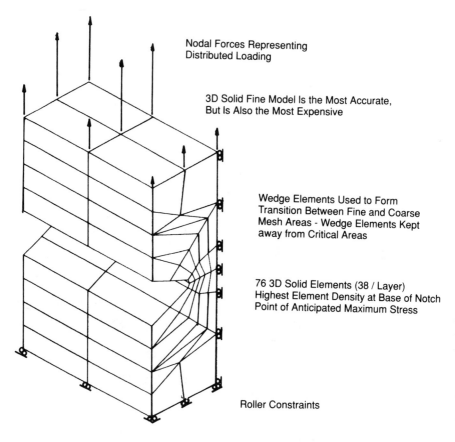

Nodal Forces Representing
Distributed Loading

3D Solid Fine Model Is the Most Accurate,
But Is Also the Most Expensive

Wedge Elements Used to Form
Transition Between Fine and Coarse
Mesh Areas - Wedge Elements Kept
away from Critical Areas

76 3D Solid Elements (38 / Layer)
Highest Element Density at Base of Notch
Point of Anticipated Maximum Stress

Roller Constraints

Figure 5.11   Notched block: Three-dimensional fine model.

5.4.3 Example: Pressure Vessel

Three separate models are formed from the same free-body diagram as shown in Figure 4.9 and discussed in Chapter 4, Sec-
ton 4.4.3. The pressure vessel is a fairly simple axisymmetric
structure. One-half of the pressure vessel is to be modeled with
two-dimensional axisymmetric elements. The loading consists of
internal, uniform pressure (p = 100.0) and the boundary constraints are roller constraints on the midplane cut boundary in

Table 5.3 Details and Computer Runtimes for Block Models

Two-dimensional
  38 Elements (two-dimensional, plane strain)
  49 Nodes 82 active DOF
  Times (sec)

Element formulation	8.149 (0.214 sec/elem.)
Wavefront solution	2.158
Stress solution	2.515 (0.066 sec/elem.)
Element forces	0.276
Misc.	3.510
Total	16.600 sec

Three-dimensional coarse
  24 Elements (three-dimensional, solid)
  57 Nodes 143 active DOF
  Times (sec)

Element formulation	34.082 (1.420 sec/elem.)
Wavefront solution	9.221
Stress solution	7.273 (0.303 sec/elem.)
Element forces	0.661
Misc.	14.43
Total	65.67 sec

Three-dimensional fine
  76 Elements (three-dimensional, solid, two layers of 38)
  249 Nodes 401 active DOF
  Times (sec)

Element formulation	106.700 (1.404 sec/elem.)
Wavefront solution	71.722
Stress solution	22.518 (0.296 sec/elem.)
Element forces	2.112
Misc.	55.052
Total	248.00 sec

Note: Computer runtimes are only valid for relative comparison between models and example cases.

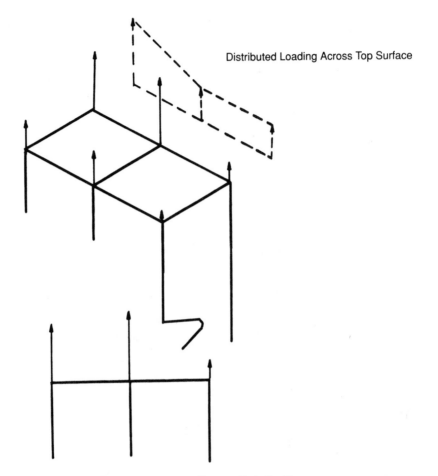

Distributed Loading Across Top Surface

Figure 5.12 Notched block: Force distribution.

the axial direction. Because this is an axisymmetric model, no displacement constraints are required in the radial direction.

The three models generated are a coarse model with 16 elements shown in Figure 5.16, a moderate model with 90 elements shown in Figure 5.17, and a fine model with 150 elements shown in Figure 5.18. Details of the three models are given in Table 5.5.

Stress results are presented in Table 5.6 and shown in Figures 5.19–5.21. Both the coarse and moderate models show the maximum principal stress to be near the midplane symmetry

Table 5.4 Stress Results of Three Notched Block Models*

	Maximum centroidal stress			Maximum nodal stress		
	σ_x	σ_y	σ_z	σ_x	σ_y	σ_z
Two-dimensional	3.24	8.70	3.58	3.40	12.68	4.82
Three-dimensional coarse	2.33	7.35	1.52	4.34	13.88	3.55
Three-dimensional fine	2.23	8.23	1.87	4.58	13.57	3.58

*σ_x, σ_y, σ_z = Normal stresses at notch

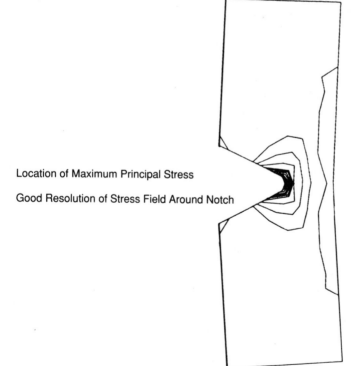

Location of Maximum Principal Stress

Good Resolution of Stress Field Around Notch

Figure 5.13 Notched block stress distribution: Two-dimensional model.

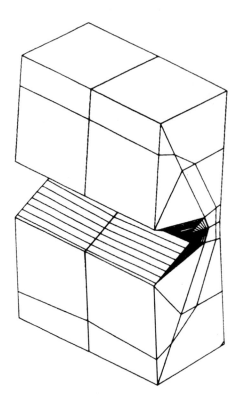

Location of Maximum Principal Stress

Poor Resolution of Stress Field at
Notch Due to Low Number of Elements

Figure 5.14 Notched block stress distribution: Three-dimensional
coarse model.

boundary. The major component of the stress at this location is
the tangential (hoop) stress. For the coarse model, the component stresses, as shown in Table 5.6, are a radial stress component of -130.6 at the surface (nodal stress) and -61.0 at the
first element centroid that represent 25% of the thickness through
the wall, an axial stress of 469.47 at the surface and 473.53 at
the centroid, and a tangential stress of 1213.35 at the surface
and 1188.1 at the centroid.

The radial stress should be equal to the applied pressure
(100.0) at the inside surface and 0 at the outside surface for
an infinitely long, uniform pressure vessel. The 130 radial

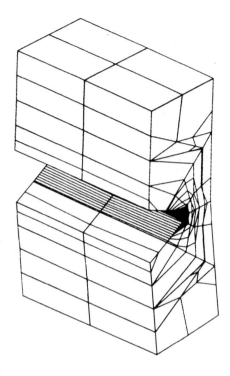

Location of Maximum Principal Stress

Good Resolution of Stress Field Around Notch

Figure 5.15 Notched block stress distribution: Three-dimensional fine model.

stress is at the inner diameter of the cut boundary and the difference between the applied pressure 100.0 and the calculated stress 130.6 is due to localized effects at the cut boundary and the imposed zero rotation. The next node away from the cut boundary node had a calculated radial stress of 81.9, indicating the steep gradient as the model attempts to achieve equilibrium. The centroidal stress of 61 is reasonable if we consider the coarseness of the mesh. The closed-form radial stress value is 75.0 at that radial location.

The axial stress is due to the pressure acting against the end of the pressure vessel and causing a tensile loading on the pressure vessel ends in the axial direction. The maximum stress occurs near the cut boundary because the wall is thinnest at this

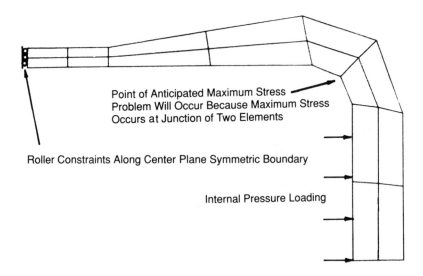

Figure 5.16 Pressure vessel: Coarse model, 16 elements.

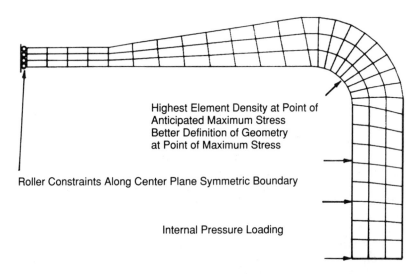

Figure 5.17 Pressure vessel: Moderate model, 90 elements.

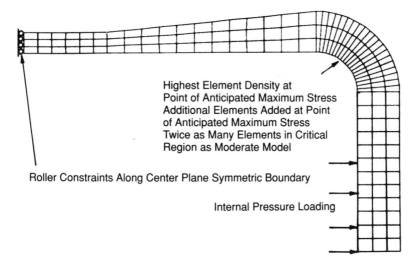

Highest Element Density at
Point of Anticipated Maximum Stress
Additional Elements Added at Point
of Anticipated Maximum Stress
Twice as Many Elements in Critical
Region as Moderate Model

Roller Constraints Along Center Plane Symmetric Boundary

Internal Pressure Loading

Figure 5.18 Pressure vessel: Fine model, 150 elements.

point and a simple stress calculation for P/A stresses gives 476.19, which compares favorably with the finite element surface axial stress of 469.47 and the centroidal stress of 473.53. No substantial difference should exist between the surface and centroid stresses because they should be constant across the section, unlike the radial stresses.

The hoop stress can be calculated from the applied internal pressure and the closed-form solution [1, 2] for an infinitely long thick-walled cylinder with a 5.00 ID and 5.50 OD with an internal pressure of 100.0 is 1052.38 at the inside diameter compared to the coarse mesh surface stress of 1213.35.

The results for the moderate mesh case are similar to those for the coarse mesh case. The surface radial stress (101.16) matches, almost exactly, the applied internal pressure (100.0). The axial centroidal stress (476.77) compares almost exactly with the P/A stresses (476.19). Both the surface (1119.55) and centroidal (1097.5) hoop stresses give better correlation with the closed-form solution of 1052.38.

The fine mesh with its better definition of the corner shows the maximum principal stress in the corner radius rather than

Table 5.5 Details and Computer Runtimes for Pressure Vessel Models

---

Coarse
   16 Elements (two-dimensional, axisymmetric)
   27 Nodes   48 active DOF
   Times (sec)
Element formulation	4.564 (0.282 sec/elem.)
Wavefront solution	1.039
Stress solution	1.733 (0.066 sec/elem.)
Element forces	0.118
Misc.	7.058

   Total   14.51 sec

Moderate
   90 Elements (two-dimensional, axisymmetric)
   127 Nodes   245 active DOF
   Times (sec)
Element formulation	25.40 (0.282 sec/elem.)
Wavefront solution	5.619
Stress solution	9.536 (0.303 sec/elem.)
Element forces	0.661
Misc.	19.80

   Total   61.02 sec

Fine
   150 Elements (two-dimensional, axisymmetric)
   204 Nodes   400 active DOF
   Times (sec)
Element formulation	42.26 (0.282 sec/elem.)
Wavefront solution	7.042
Stress solution	15.95 (0.296 sec/elem.)
Element forces	1.145
Misc.	28.11

   Total   94.515 sec

Note: Computer runtimes are only valid for relative comparison between models and example cases.

Table 5.6 Stress Results of Three Pressure Vessel Models

Maximum centroidal stress				Maximum nodal stress			
σ_r	σ_a	σ_t	σ_1	σ_r	σ_a	σ_t	σ_1
Coarse							
-61.019	473.53	1188.1	1188.1	-130.6	469.47	1213.35	1213.35
Moderate							
-81.061	476.77	1097.5	1097.5	-101.16	479.56	1119.55	1119.55
Fine							
594.18	565.04	235.47	1096.6	943.13	991.78	436.95	1836.16

σ_r = radial stress
σ_a = axial stress
σ_t = tangential (hoop) stress
σ_1 = maximum principle stress

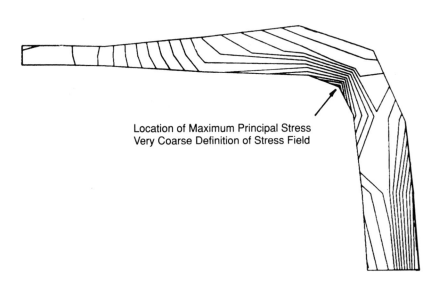

Location of Maximum Principal Stress
Very Coarse Definition of Stress Field

Figure 5.19 Pressure vessel stress distribution: Coarse model.

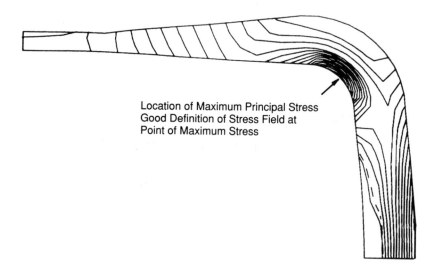

Location of Maximum Principal Stress
Good Definition of Stress Field at
Point of Maximum Stress

Figure 5.20 Pressure vessel stress distribution: Moderate model.

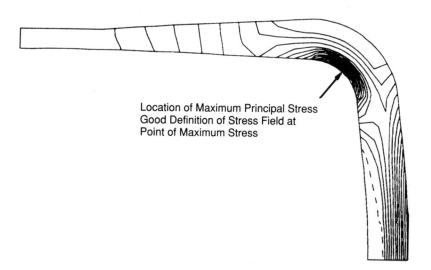

Location of Maximum Principal Stress
Good Definition of Stress Field at
Point of Maximum Stress

Figure 5.21 Pressure vessel stress distribution: Fine model.

near the cut boundary where the wall thickness is the thinnest. For the fine mesh model, the results near the cut boundary are still consistent with those for the coarse and moderate models, with the element centroidal stresses $\sigma_r = -81.392$, $\sigma_a = 486.12$, $\sigma_t = 1081.5$, and $\sigma_1 = 1081.5$.

This model, although simple in appearance, has a complex, three-dimensional stress field with individual component stresses, as discussed above, resulting from several phenomena. Because the maximum principal stress is used as the criterion for reporting stresses in Table 5.6, a change in one of the component stresses caused enough of a change in the principal stresses to change the location of the maximum principal stress even though there is no substantial difference in the principal stresses at the cut boundary location between the three cases (coarse $\sigma_1 = 1188.1$, moderate $\sigma_1 = 1097.5$, and fine $\sigma_1 = 1081.5$). At the corner location, the component stresses in Table 5.6 for the fine mesh case show almost equal radial and axial stress components, which is to be expected because it is intuitively obvious that the maximum stress will be parallel to the inside surface of the pressure vessel.

This pressure vessel is a geometrically simple structure with simple loading and boundary conditions. It has, however, an interesting stress distribution that may be easily understood and checked by using the basic principles of solid mechanics. Refinements to the mesh cause some changes in the calculated results at the cost of computer time, as shown in Table 5.5. The comparison of computer runtimes for the various models cannot be done on a percentage basis alone, i.e., the coarse model runs in 15% of the time required for the fine model, but the absolute runtimes must be taken into account. For this case, the fine model required only 150 two-dimensional elements to give a reasonably detailed mesh. There is little reason to go to a coarse mesh because the computer time and cost will be relatively small compared to the engineering effort required to set up the model and analyze the results.

5.4.4 Example: Crank

At the free-body diagram step as discussed in Section 4.4.4, it is decided to include only the crank itself in the finite element model. The free-body diagram of the crank showing the displacement constraints at the shaft bore and keyway location and the applied loading at the handle bore is shown in Figure 4.10.

The crank is modeled with three-dimensional eight-node solid elements. The gridwork is shown in Figure 5.22. One layer of elements is used to model the flange connecting the shaft hub to the handle hub. One additional layer of elements is used to model the shaft hub and a third layer to model the handle hub. The model that is developed is suitable to give an overall representation of the stresses in the crank without concentrating on any one area.

It may be anticipated that the highest stresses may occur at the corners of the keyway, and if that is the case, a finer grid may give a better resolution of the stress field and a better estimate of the peak stress. A method for accomplishing this is refined mesh modeling, which is discussed in Chapter 9. With this technique, a second model of the keyway region is formed and displacements from the complete model corresponding to the outside boundaries of the refined model are transferred to the refined model. The difficulty in including a fine mesh near the keyway in the complete model is that it requires more total elements in the model and several layers of transition elements to

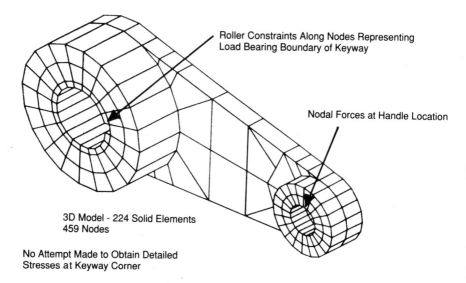

Roller Constraints Along Nodes Representing
Load Bearing Boundary of Keyway

Nodal Forces at Handle Location

3D Model - 224 Solid Elements
459 Nodes

No Attempt Made to Obtain Detailed
Stresses at Keyway Corner

Figure 5.22 Crank model.

Pattern of Triangular Elements Used for Transition
Reduction of 2:1 for Each Transition Layer

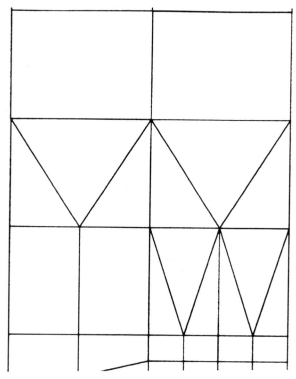

Figure 5.23 Coarse to fine mesh transition.

reduce from the fine mesh to the coarse mesh, as shown in Figure
5.23.

A second area of potentially high stress that may be addressed
with a refined mesh model is the welded interface of the shaft hub
and the flange. This model does not attempt to represent the weld
fillet and assumes homogeneous material. The model would be
equally applicable to a crank made as a weldment or casting.
Again, additional detail could be added to the interface at the
expense of driving up the computer runtime. The computer

Table 5.7 Details and Computer Runtimes for Crank Model

Three-dimensional	
224 Elements (three-dimensional, solid)	
459 Nodes 1306 active DOF	
Times (sec)	
Element formulation	320.53 (1.431 sec/elem.)
Wavefront solution	427.95
Stress solution	77.609 (0.346 sec/elem.)
Element forces	5.821
Misc.	37.418
Total	924.38 sec

Note: Computer runtimes are only valid for relative comparison between models and example cases.

runtime breakdown for this model is given in Table 5.7. This fairly uniform model is constructed without attempting to presuppose the exact location of maximum stress. In this way, the model can give an estimate of the stress distributions and provide data that may be used in subsequent refined model runs.

A plot of maximum principal stress results is shown in Figure 5.24. The results show two areas of high stress, the keyway and the interface of the shaft hub and flange. The maximum stresses for each location are given in Table 5.8.

5.4.5 Example: Gear Segment

Three different models are used to calculate the stresses in the gear segment with particular emphasis on the fillet area where fatigue cracks are known to develop, as discussed in Chapter 4, Section 4.2.5. The three models are the following.

1. *Two-dimensional single tooth model* Shown in Figure 5.25, this model uses roller constraints along its cut boundaries between the teeth and roller constraints along the base of the segment. The loading is applied directly to the tooth pressure surface based on the average force per length of the most heavily loaded section of length of the tooth. Figure 5.26 shows a pressure distribution across three teeth in contact. The two-dimensional single tooth model represents a tooth as it is taking on its maximum loading.

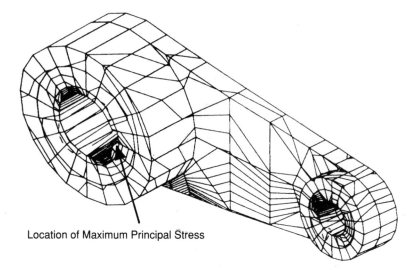

Location of Maximum Principal Stress

Figure 5.24 Crank: Maximum principal stress distribution.

Table 5.8 Stress Results of Crank Model

Maximum centroidal stress				Maximum nodal stress			
Keyway							
σ_r	σ_t	σ_a	σ_1	σ_r	σ_t	σ_a	σ_1
-371.38	785.90	66.87	1126.8	-538.41	814.92	30.00	1085.19
Hub-flange interface							
σ_x	σ_y	σ_z	σ_1	σ_x	σ_y	σ_z	σ_1
1150.9	78.62	3.96	1151.4	986.81	127.43	-5.16	1014.06

σ_r = radial stress at shaft bore, near keyway
σ_t = tangential stress at shaft bore, near keyway
σ_a = axial stress at shaft bore, near keyway
σ_1 = maximum principal stress, near keyway
σ_x = X direction stress, X axis is a line connection the two bores
σ_y = Y direction stress, Y axis is in the plane of the flange, nor-
       mal to the X axis
σ_z = Z direction stress, Z axis is normal to the plane of the flange,
       in the same direction as the axial direction above
σ_1 = maximum principal stress

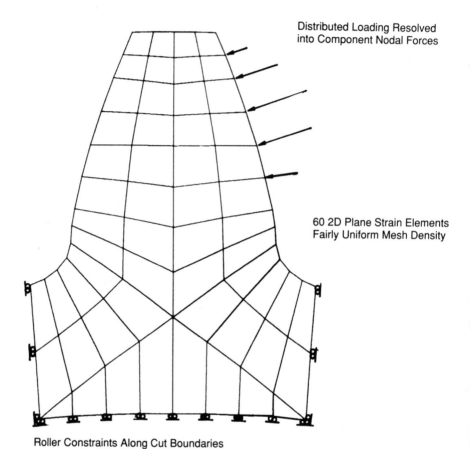

Distributed Loading Resolved
into Component Nodal Forces

60 2D Plane Strain Elements
Fairly Uniform Mesh Density

Roller Constraints Along Cut Boundaries

Figure 5.25 Two-dimensional single gear tooth model.

2. *Two-dimensional three-tooth segment model* Shown in Figure 5.27, this model is simply the single tooth model repeated three times. Loadings are applied to the profile in the same manner as for the single tooth model. Roller displacement constraints are used along the cut boundaries and the bottom of the segment. Loadings are applied to all three teeth in a manner that represents the instant at which the maximum loading is on the center tooth.

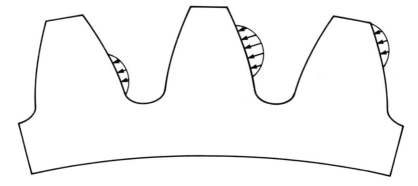

Note: Difference in Loading Between Teeth (Both Location and Magnitude)
due to Angular Position of Teeth Relation to Mating Gear Teeth

Figure 5.26 Pressure distribution across three teeth.

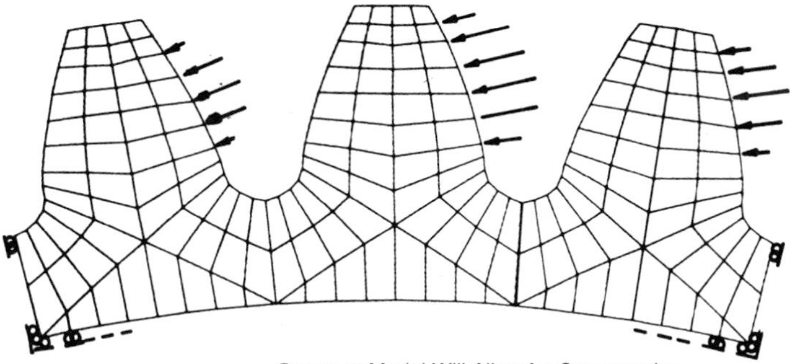

Distributed Loading Resolved into Component Nodal Forces

Roller Constraints Along Cut Boundaries

Segment Model Will Allow for Compressive
Stress Effects to Be Calculated
180 2D Plane Strain Elements
Same Mesh Pattern as Single Tooth
Model But Repeated to Form Segment

Figure 5.27 Two-dimensional three-tooth segment model.

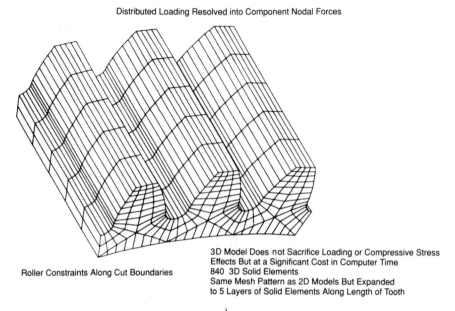

Figure 5.28 Three-dimensional three-tooth segment model.

3. *Three-dimensional three-tooth model* Shown in Figure 5.28, this model has the same gridwork across its face as the two-dimensional models. However, it has five layers of three-dimensional elements along the length. The applied loading is a pressure distribution across the pressure surface of each of the three teeth. The pressure distribution varies across the profile as did the two-dimensional pressure distribution, and it also varies along the length of the teeth, a factor that the two-dimensional model could not represent.

Stress results for the two-dimensional three-tooth model should be more accurate than the results from the two-dimensional single tooth model. In any pair of gears, several teeth carry the load at any one time. This load sharing affects fillet stresses not only from loading of a particular tooth, but also in obtaining an accurate stress profile in the fillet. As a tooth begins to come into contact, there is loading on the preceding tooth, and there may be a compressive stress in the fillet due to the bending of the preceding tooth. This effect is described by Wilcox [3] for spiral bevel gears, but it holds for spur gears as well. For the purposes of a subsequent fatigue analysis, the total stress range is the significant parameter. Therefore, it is important to quantify this negative portion of the stress cycle in order to accurately determine the cyclic stress range.

The results of the two models are given in Table 5.9. This table shows the maximum stresses that occur at the same location on tooth number 2 of the two-dimensional three-tooth model as on the two-dimensional single tooth model. Stresses in the two-dimensional three-tooth model are slightly lower than those for the two-dimensional single tooth model probably because of the compressive stress contribution due to the loading of tooth number 1. The comparison of computer runtimes in Table 5.10 shows that, as expected, the two-dimensional three-tooth model required just over three times as much computer time as the two-dimensional single tooth model. The times for element formulation and element stress recovery are linear, with the number of elements and average time per element for these two functions fairly close. Theoretically, there should be no difference in the per element

Table 5.9 Stress Results of Three Gear Models

Maximum centroidal stress				Maximum nodal stress			
σ_r	σ_a	σ_t	σ_1	σ_r	σ_a	σ_t	σ_1
Two-dimensional single-tooth, plane strain							
506.7	228.4	254.4	588.0	904.3	382.8	371.7	1063.2
Two-dimensional three-tooth model, plane strain							
465.4	234.3	315.5	632.9	833.4	417.2	557.2	1173.4
Three-dimensional three-tooth model							
280.8	99.2	228.4	385.2	514.7	179.2	326.6	714.3

σ_r = radial stress
σ_a = axial stress
σ_t = tangential stress
σ_1 = maximum principle stress

Table 5.10 Details and Computer Runtimes for Gear Models

Two-dimensional single tooth		
60 Elements (two-dimensional, plane strain)		
79 Nodes 135 active DOF		
Times (sec)		
Element formulation	13.14	(0.219 sec/elem.)
Wavefront solution	3.767	
Stress solution	3.955	(0.066 sec/elem.)
Element forces	0.445	
Misc.	13.03	
Total	34.337 sec	

Two-dimensional three-tooth segment		
168 Elements (two-dimensional, plane strain)		
211 Nodes 421 active DOF		
Times (sec)		
Element formulation	39.03	(0.232 sec/elem.)
Wavefront solution	21.47	
Stress solution	17.05	(0.101 sec/elem.)
Element forces	12.46	
Misc.	30.83	
Total	120.83 sec	

Three-dimensional three-tooth segment		
840 Elements (three-dimensional, solid)		
1270 Nodes 3808 active DOF		
Times (sec)		
Element formulation	1198.09	(1.426 sec/elem.)
Wavefront solution	3332.49	
Stress solution	250.748	(0.299 sec/elem.)
Element forces	297.542	
Misc.	301.192	
Total	5380.07 sec	

Note: Computer runtimes are only valid for relative comparison between models and example cases.

time for element formulation and stress recovery. A technique for minimizing computer time for repetitive structures such as the gear teeth is substructuring, which is discussed in more detail in Chapter 10.

Plots of maximum principal stresses are shown in Figure 5.29 for the two-dimensional single tooth model and in Figure 5.30 for the two-dimensional three-tooth model.

Location of Maximum Principal Stress
Maximum Tensile Bending Stress Occurs
in Fillet as Anticipated

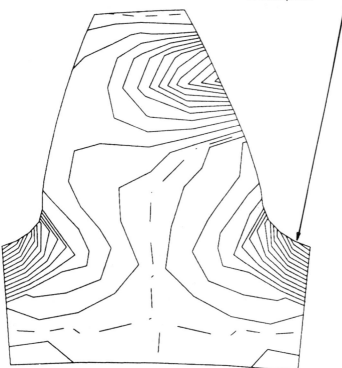

Figure 5.29 Two-dimensional single tooth model stress distribution.

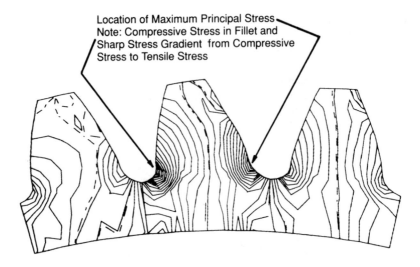

Location of Maximum Principal Stress
Note: Compressive Stress in Fillet and
Sharp Stress Gradient from Compressive
Stress to Tensile Stress

Figure 5.30 Two-dimensional three-tooth model stress distribution.

In order to fully utilize the two-dimensional three-tooth model, there should be a series of static runs with the loading changed between and over the teeth to simulate the teeth, and especially tooth number 2, as they roll through contact. The results of this series of runs, when assembled, provide a pseudodynamic stress time history for a tooth. Required input to such a series of runs would be a description of the force distribution on the contacting teeth as a function of angular position (time). This drawback could be eliminated by expanding the two-dimensional model to include three teeth from the driving gear as well as the driven gear and using gap elements to represent the contact between the teeth.

The three-dimensional three-tooth model uses the same element density per plane as the two-dimensional three-tooth model. The three-dimensional model uses five layers of elements along the length of the teeth as shown in Figure 5.28. A stress contour plot for the three-dimensional model is shown in Figure 5.31.

The three-dimensional model has three basic advantages over the two-dimensional models: (1) the three-dimensional models calculate the full three-dimensional stress field, the two-dimensional models cannot calculate an axial stress or associated shear

3D Model Shows Elliptical Loading Areas on Pressure Faces

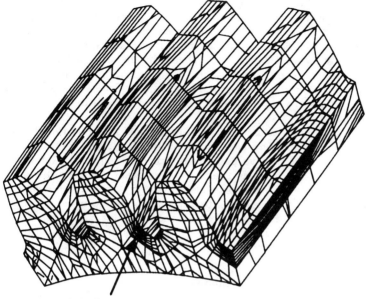

Location of Maximum Principal Stress
Lengthwise Variation in Stress Follows Variation in Loading
on Any Given Plane Pattern of Stress Follows 2D Pattern

Figure 5.31 Three-dimensional three-tooth model stress distribution.

stresses; (2) the three-dimensional model can account for a variation in contact pressure along the length of the tooth; (3) the three-dimensional model, if used together with a gear segment model of the mating gear, can include lengthwise "crowning" of the teeth to accurately evaluate its effect on contact pressure. Lengthwise crowning is used to localize the contact pattern into a well-defined ellipse that is less susceptible to deflections of the gears and their bearings. Without crowning, there will be full contact along the length of the teeth under perfect alignment conditions, but highly localized line contact under less than perfect conditions.

A comparison of stresses between the two-dimensional three-tooth model and the three-dimensional three-tooth model shows

that the location of the maximum stress occurs at approximately
the same location in the fillet. The magnitude of the stresses in
the three component directions is about of the same magnitude.
There are two reasons for the differences in stress: (1) The
two-dimensional model uses a plane strain calculation to obtain
the axial stresses, whereas the three-dimensional model calculates these stresses directly, and (2) the loading of the three-
dimensional model is distributed over the length of the gear tooth
rather than applying an average force per length in the two-dimensional model.

The more sophisticated three-dimensional analysis is not without its price. The three-dimensional model with its 840 solid
elements and 3808 active degrees of freedom required 44.5 times
as much computer time as the two-dimensional three-tooth model
with only 168 two-dimensional elements and 421 active degrees
of freedom. The decision as to whether or not to use a three-
dimensional model is obviously not made without a good understanding of the requirements of the analysis and the operating
conditions of the gear set.

5.4.6 Example: Turbine Blade

A three-dimensional model using eight-node solid elements is developed from the turbine blade and shown in Figure 5.32. A portion of the cover is included in the model to properly apply centrifugal loading to the tenon, as described in the discussion of
the free-body diagram. A segment of the rotor around the blade
root attachment is also included in the model, as shown in the
free-body diagram (Figure 4.13).

Attachment between the tenon and cover is made by coupling
all three degrees of freedom at the appropriate nodes. In the
blade root-rotor attachment, the nodes along the six contact surfaces are coupled. No attempt is made to use gap elements to
represent the interface as they require a nonlinear, iterative solution and would significantly increase the computer time.

A total of 774 elements is used in the model. This allowed for
a reasonable representation of the airfoil section with two elements
across the thickness, eight elements across the chord, and 18-ele-
ment layers along the length (Figure 5.33). Accurate geometric
modeling of the airfoil geometry is more important in the calculation of natural frequencies and dynamic response than in the calculation of static stress.

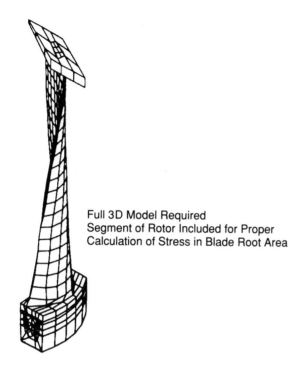

Full 3D Model Required
Segment of Rotor Included for Proper
Calculation of Stress in Blade Root Area

Figure 5.32 Turbine blade model.

Six elements are used to represent the tenon and eight elements are used in the cover (Figure 5.34). This is sufficient
to apply the centrifugal loading to the airfoil tip, but not really
adequate to model the stresses at the tenon-cover interface or
at the fillet between the tenon and airfoil tip. Although the
blades are connected by the cover into groups of three, a simplifying assumption is made that under static loading, all three
blades in the group will see approximately the same stress.
Therefore, a single blade with one cover segment is used with
constraints on the sides of the cover segment to prevent rotation.

A total of 335 elements are used in the combined blade and
rotor root attachment area. The details of the blade root portion

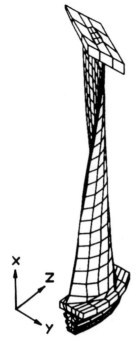

Rotor Segment Removed to Show Model of Blade Root

Figure 5.33 Turbine blade model without rotor segment.

of the model are shown in Figure 5.35 and the rotor slot model is shown in Figure 5.36. The nodes in the blade root portion of the model are separate from the nodes of the rotor slot portion. The contact between the blade root and rotor slot is modeled by coupling the coincident pairs of nodes on the contact surfaces. The root cross section is capable of giving a reasonable approximation of the highly localized stresses in that area. A refined two-dimensional model may be used to give a better resolution of the peak stresses, as discussed in Chapter 9.

Computer runtimes for the three-dimensional model are given in Table 5.11. Stress results for the various components of the model are given in Table 5.12. Plots of maximum principal stress

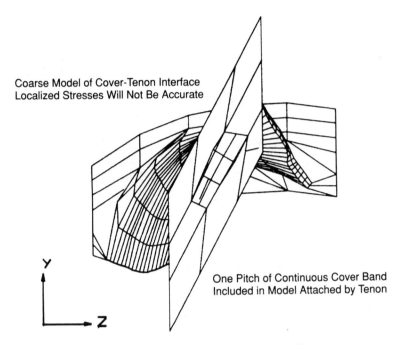

Coarse Model of Cover-Tenon Interface
Localized Stresses Will Not Be Accurate

Y

One Pitch of Continuous Cover Band
Included in Model Attached by Tenon

Z

Figure 5.34 Turbine blade model: Cover details.

distributions are given in Figures 5.37 for the airfoil section and
5.38 for the root section.

5.5 SUMMARY

Developing the most cost-effective model requires the combination
of knowledge of finite element principles, behavior of the structure to be analyzed, and engineering experience and judgment.
The process that some engineers consider to be an art can, however, be made into more of a science by taking a step-by-step
approach and isolating the critical gridwork layout step. Probably, the most critical item in determing the number and placement of elements within the gridwork is understanding of the behavior of the selected element. The following chapter gives data
and some discussion into the properties of some common elements.

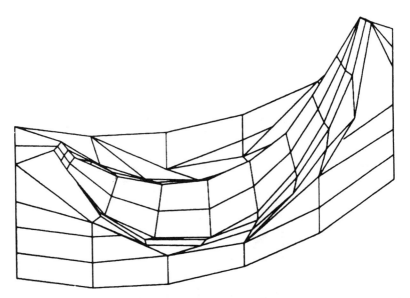

Note: Complex Transition Between Airfoil and Platform Sections
Virtually Impossible to Avoid. Distorted Elements in Transition
Stresses in These Elements Should Be Ignored

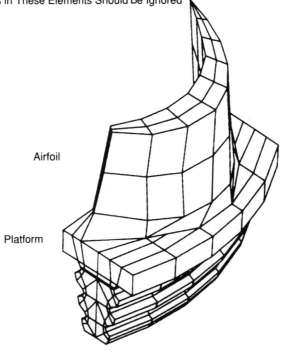

Airfoil

Platform

Figure 5.35 Turbine blade model: Blade root details.

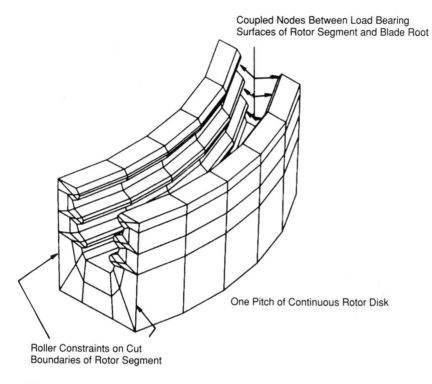

Coupled Nodes Between Load Bearing
Surfaces of Rotor Segment and Blade Root

One Pitch of Continuous Rotor Disk

Roller Constraints on Cut
Boundaries of Rotor Segment

Figure 5.36 Turbine blade model: Rotor slot details.

Table 5.11 Details and Computer Runtimes for Turbine Blade Model

774 Elements (three-dimensional, solid)
1184 Nodes   3302 active DOF
Times (sec)

Element formulation	1106.82	(1.430 sec/elem.)
Wavefront solution	3132.66	
Stress solution	229.878	(0.297 sec/elem.)
Element forces	273.302	
Misc.	288.652	
Total	5031.31	sec

Note: Computer runtimes are only valid for relative comparison
between models and example cases.

Table 5.12 Stress Results of Turbine Blade Model

	Maximum centroidal stress			
	σ_r	σ_a	σ_t	σ_1
Cover	6317.2	16171.0	7656.1	30474.0
Tenon	37560.0	2869.6	-6300.2	40353.0
Airfoil	59789.0	2177.9	1584.4	61665.0
Root attachment	95164.0	33315.0	43192.0	96797.0

σ_r = radial stress.
σ_a = axial stress.
σ_t = tangential stress.
σ_1 = maximum principal stress.

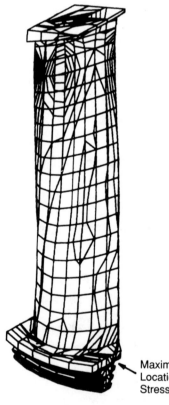

Maximum Principal Stresses
Location of Maximum Principal
Stress in Root Notch

Figure 5.37 Turbine blade model stress distribution.

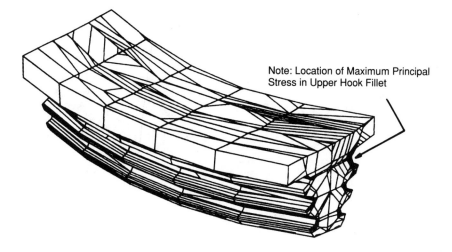

Note: Location of Maximum Principal
Stress in Upper Hook Fillet

Figure 5.38 Turbine blade root principal stress distribution.

REFERENCES

1. Seeley, F. B. and J. O. Smith, *Advanced Mechanics of Materials*, 2nd ed., Wiley, New York, 1932, pp. 304–306.
2. Timoshenko, S., *Strength of Materials — Part II*, 3rd ed., Van Nostrand, New York, pp. 205–214.
3. Wilcox, L. E., "An Exact Analytical Method for Calculating Stresses in Bevel and Hypoid Gear Teeth," Proceedings of International Symposium on Gearing and Power Transmission, Tokyo, Japan, 1981.

6

DEBUGGING FINITE ELEMENT MODELS

A large proportion of the total time in generating a finite element model is spent in debugging the model. Prior to the use of interactive mesh generation programs, the time spent in model debugging, correction, and input data editing (either card decks or batch input files) could exceed the time for actual data input by a factor of 2 or 3. The use of interactive graphic mesh generation software has had its greatest impact on cutting model debugging time by allowing the user to view each part of the model as it is input and make corrections immediately, when necessary. Even with the older batch-type input, models could be developed with a minimum number of input statements; however, there was always a tendency to build an entire model before submitting a check run to look for errors. Error location was difficult because a plot of the entire model may reveal several problems that would be hard to pinpoint.

A typical finite element model can be divided into four areas:

Geometry.
Material properties.

Table 6.1 Common Symptoms and Their Possible Causes

Excessive deflection, but anticipated stress	Young's modulus too low, missing nodal constraints
Excessive deflection and excessive stress	Applied force too high, nodal coordinates incorrect, force applied at wrong nodes
Internal discontinuity in stress and deflection	Force applied at wrong node, missing or double internal element
Discontinuity along boundary	Missing nodal constraint, force applied at wrong node
Higher or lower natural frequencies than anticipated	
Static deflections, O.K.	Density incorrect
Static deflections, not O.K.	Young's modulus incorrect
Internal gap opening up in model under load, stress discontinuity	Improper nodal coupling

Applied forces.
Displacement constraints.

Some of the more common problems in these areas are given in Table 6.1 and discussed in the following sections.

6.1 GEOMETRY

Geometry problems can be classified as obvious (element distortion) and not so obvious (unconnected or missing elements, slight inaccuracies in dimensions). The obvious problems stem from the nodal coordinates and/or element descriptions that result in badly distorted elements in an element plot.

When data are input interactively, it is recommended that plots be made frequently, i.e., as each set of nodes or elements are either input or generated (Figures 6.1 and 6.2).

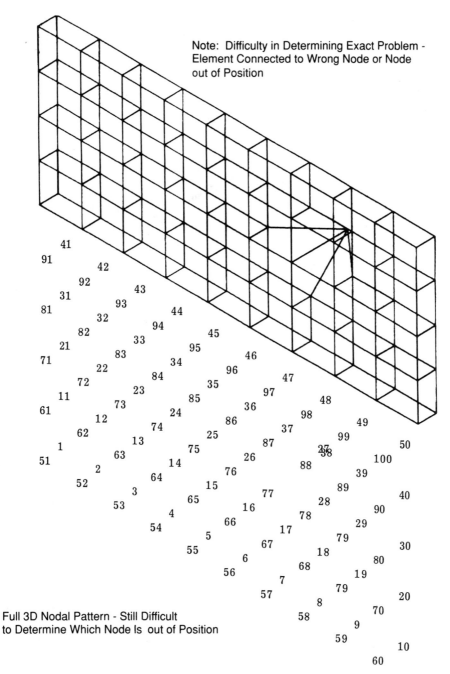

Figure 6.1 Simple note and element pattern with single nodal error.

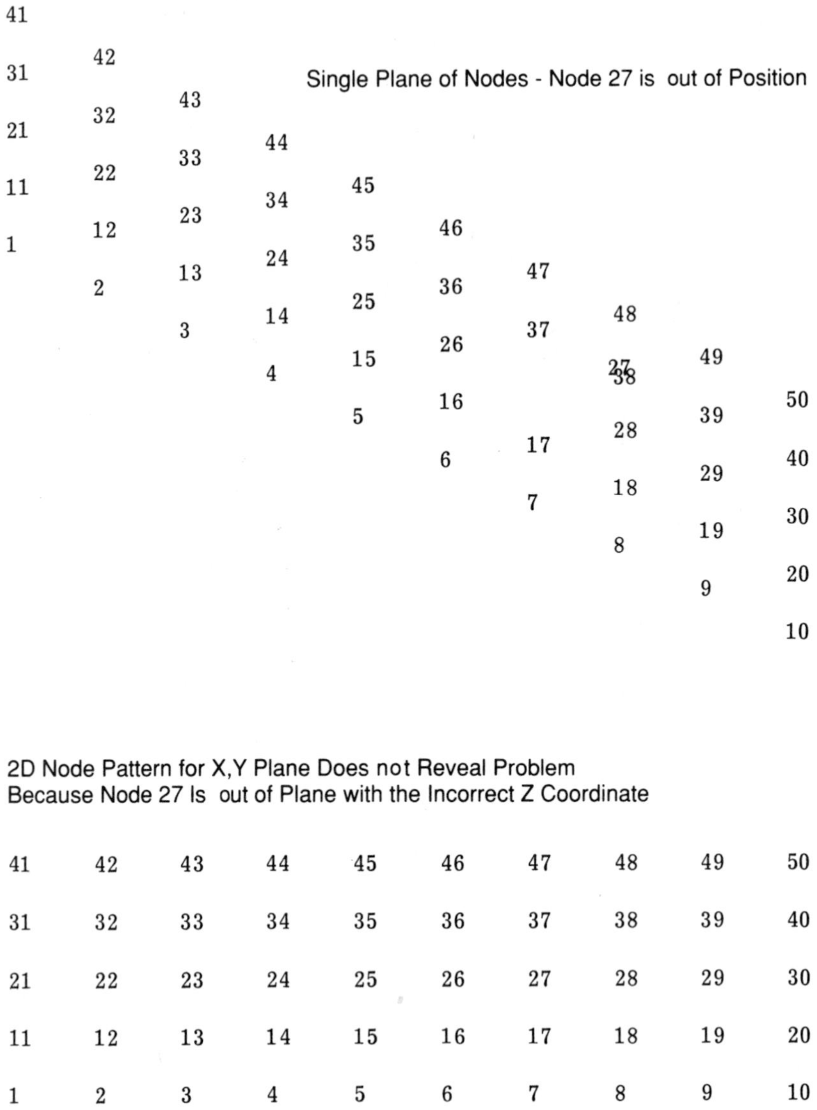

41

42

31

43

32

21

33

22

44

11

34

45

23

46

12

35

1

24

47

13

25

36

2

48

14

26

37

3

15

49

4

27
38

50

5

16

39

28

40

6

17

29

18

30

7

19

8

20

9

10

Single Plane of Nodes - Node 27 is out of Position

2D Node Pattern for X,Y Plane Does not Reveal Problem
Because Node 27 Is out of Plane with the Incorrect Z Coordinate

41	42	43	44	45	46	47	48	49	50
31	32	33	34	35	36	37	38	39	40
21	22	23	24	25	26	27	28	29	30
11	12	13	14	15	16	17	18	19	20
1	2	3	4	5	6	7	8	9	10

Figure 6.2 Three-dimensional nodal pattern plot slowing error
in nodal location.

Nodes are typically input first and plots of nodal patterns should be made at each major step in the generation process. Once the user is satisfied that the nodal pattern is correct, the element connectivity may be entered. Then if elements appear to be improper, the user knows that it is an error in the element numbering. If no plots are made until the elements are entered, the user does not know whether the error is in the nodes or elements and the user must "back track" to find the error.

When a complex model is being generated, it is advisable to enter nodes and elements in sections rather than try to enter all the nodes and, then, all the elements. The rule of thumb should be not to enter more data at a time than can be verified by one or two plots. It is much easier to fix problems as they occur than to let an error slip through that may not be noticed.

Not so obvious element geometry problems are missing or overlapping elements: coupled nodes and small errors in nodal coordinates. Missing elements can most often be discovered by using a "shrink" option for element plotting as shown in Figure 6.3.

One or more missing internal elements may be difficult to pick up if all adjacent elements are present. By the same token, it is possible to define double elements in the same space attached to the same nodes. The finite-element program will accept this and the result will be double stiffness, or a hard spot, in the model that will be difficult to determine without a thorough examination of the results. The dangerous aspect of missing or double elements is that they probably do not affect the results enough to be obvious, but could lead to inaccuracies in the range of 5–10%.

When complex models are developed, they are sometimes generated in sections with coincident, but uniquely defined nodes along their common boundary. The intention is to couple these coincident nodes; however, this must be checked by a printout of a check run because an element plot typically would not reveal a problem because the nodes are coincident. If graphics software is available that can highlight free surfaces, this may reveal any internal gaps in the model.

Another area that can lead to inaccuracies is nodal coordinate errors, which cannot be picked up by a visual inspection. An error of 10% on a dimension of a model may not be obvious in an element plot, but will be significant in the final results. When constructing a model, it is advisable to compare the model coordinates against the structure's blueprint. This involves

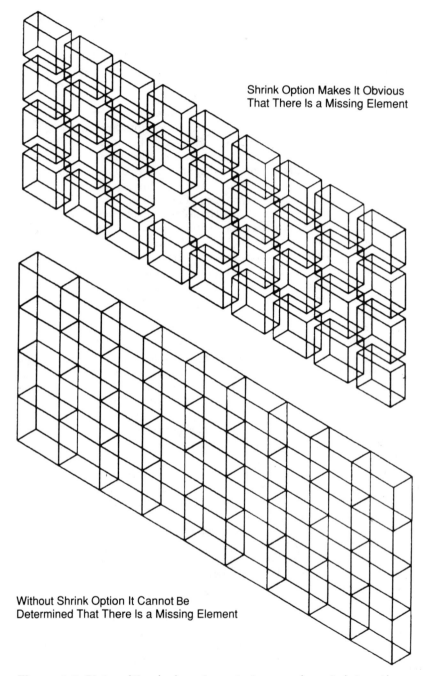

Shrink Option Makes It Obvious
That There Is a Missing Element

Without Shrink Option It Cannot Be
Determined That There Is a Missing Element

Figure 6.3 Plate with missing element shown using shrink option.

extracting nodal coordinates of key points on the model and making the appropriate subtractions to compare to the linear dimensions of the blueprint.

6.2 MATERIAL PROPERTIES

Data input for material properties is minimal: Young's modulus and Poisson's ratio are always required. For dynamic models, mass density is also required. For thermal models, specific heat, conductivity, and boundary data such as film coefficients or radiation view factors may also be required. Material properties must be checked by having the program echo back the input data either to a terminal or in a printout. The user should never assume that he entered the data correctly even if it seems simple. This print-back takes only a few seconds, but must be checked carefully to avoid embarrassing mistakes later in the analysis.

The two most common mistakes are

1. Young's modulus is incorrect by a power of 10.
2. Inconsistent units are assumed for material density. Because the finite element method is nondimensional, the user must insure that the units assumed for density are consistent with the units used for stiffness, time, force, and acceleration in a dynamic analysis.

When more than one material is used in a model, some element plot routines can indicate the different materials with different colors for checking.

6.3 APPLIED FORCES

Forces must be applied at the nodes in (1) the proper magnitude, (2) the proper location, and (3) the proper direction. Forces may be checked either from the print-back or from a node or element plot with the appropriate force display option selected.

In checking to see if the forces are applied at the correct location, both the node number and the location of the node must be checked, i.e., the node number may be correct but if the nodal coordinates are not correct, the force will not be applied at the correct location. Element or node plots with a force boundary condition option selected will show the locations and directions of the applied forces. The actual force magnitude must be checked from a print-back, either from a terminal or print-out.

A nodal map, i.e., a plot of the node locations, is a great help in assigning and checking nodal force loadings.

As in the case of material properties, care must be taken to insure that the units selected are consistent with the nodal co-ordinates and material properties. One parameter that can cause problems is rotational speed when centrifugal force is applied. To be consistent with the equation, units must appear in radians per second (rad/sec):

$$F = m \ r \ \omega^2$$

where

F = force (lbf)

m = mass $\dfrac{\text{lbf}}{\text{in./sec}^2}$

r^2 = length, radius (in.)

ω = rad/sec

6.4 DISPLACEMENT CONSTRAINTS

Checking of displacement constraints is similar to checking ap-plied forces. Generally, nodes are constrained with a displace-ment of 0, which leaves only proper location and direction. Here again, a node map and the use of a node or element plot with the boundary condition option selected are very useful.

Two common problems with displacement boundary conditions are (1) missing constrained nodes and (2) an overconstrained boundary. When a series of nodes is representing a uniformly constrained boundary, a missing nodal constraint can lead to a local stress discontinuity and overall higher displacements. When specifying constraints along a symmetry boundary, care must be taken not to overconstrain the boundary by specifying more de-grees of freedom per node than are appropriate.

6.5 SUMMARY

The use of interactive computer graphics has made possibly its strongest contribution to finite-element analysis productivity in assisting the user to debug models. The efficiency of an analy-sis must be measured in terms of net time for the analysis, which is the total time from the definition of the problem to the point of obtaining useful results. A good checklist is invaluable in working out finite element model problems. No piece of data should be taken casually or assumed to be correct. Experienced users check even the most simple input data when a problem is encountered. More often than not, the problem is something embarrassingly simple.

7

ELEMENT PERFORMANCE

In forming a finite element model, probably the most critical decision is the type of elements to be used. The various types of elements each have a particular geometry and characteristics that allow them to model certain "theoretical" shapes with nearly perfect accuracy. For example, plate elements will model a simple plate in tension or bending with a minimal number of elements, and a single element will suffice if the structure is simple enough.

Actual structures are hardly ever so simple that they can be perfectly represented by only a few elements (if they were, there would be no need for the finite element method). In real cases, the engineer must make his decision on which elements to use to model his structure. Following that, he needs to determine the absolute and relative element densities to use to achieve the desired accuracy with an acceptable computer runtime.

Chapter 2 discussed the various types of elements and their mathematical formulations. The main element categories are

Two-dimensional, plane stress/plane strain/ axisymmetric
Three-dimensional, solid
Plate and beam

Within each general category there are parameters that differen-
tiate the element's behavior. For two- and three-dimensional
solid elements, the main differences are the order of integration,
method of formulation of the stiffness matrix, method of resolution of mass, and number of integration points used for Gaussian
quadrature integration. The most common element types, classified according to their geometry, are given in Table 7.1 and
shown in Figures 7.1–7.5.

Table 7.1 Classes of Element Geometry Types

I   Two-dimensional plane stress, plane strain, axisymmetric
      Quadrilateral or triangle
      2 DOF per node
      Ref.: Chapter 2, Section 2.5–2.6
      Figure 7.1

II   Three-dimensional solid
      Hexahedron (brick), wedge, or tetrahedron
      3 DOF per node
      Ref.: Chapter 2, Section 2.8
      Figure 7.2

III   Plate
      Quadrilateral or triangle
      6 DOF per node
      Ref.: Chapter 2, Section 2.8
      Figure 7.3

IV   Beam
      No shape, two nodes representing end points; cross-sec-
      tional properties specified directly
      6 DOF per node
      Ref.: Chapter 2, Section 2.8
      Figure 7.4

V.   Specialty elements
      Point mass
      Spring
      Damper
      Gap
      Figure 7.5

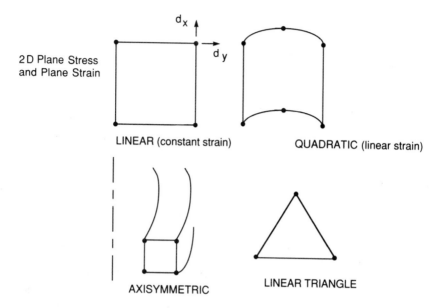

2 D Plane Stress
and Plane Strain

LINEAR (constant strain) QUADRATIC (linear strain)

AXISYMMETRIC LINEAR TRIANGLE

Figure 7.1 Two-dimensional elements; three or four nodes; 2
DOF per node.

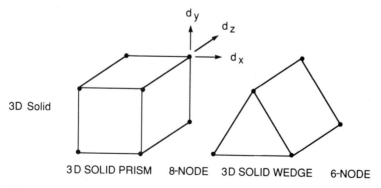

3D Solid

3D SOLID PRISM 8-NODE 3D SOLID WEDGE 6-NODE

Figure 7.2 Three-dimensional solid elements: 3 DOF per node.

Plate and Beam

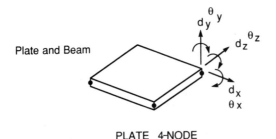

PLATE 4-NODE

Figure 7.3 Plate elements: three or four nodes; 6 DOF per node.

The accuracy of any finite element model is strongly dependent on the type of elements used in the model, as well as the number of elements, their relative distribution, the type of loading, and the amount of element distortion. Table 7.2 lists the main parameters that affect element accuracy. There is no simply defined relationship between the parameters listed in Table 7.2, and it would be wrong to assume that net accuracy (or error) would simply be a linear combination of some value assigned to each of the parameters. In addition to individual element accuracy, the net accuracy of a model is a function of how well the selected elements match the structural geometry, e.g., straight-sided elements approximating a curved boundary.

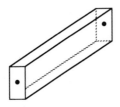

BEAM 2-NODE

Figure 7.4 Beam elements: 6 DOF per node.

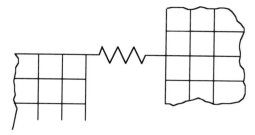

Spring Connects Two Nodes with a Single DOF Stiffness
More Than One Spring May Be Used to Connect Nodes with
Multiple DOF

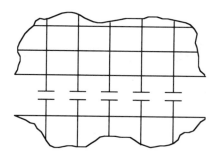

Gap High Stiffness when Closed (10x Stiffness of Surrounding
Elements)
Zero Stiffness when Open
Allows for Sliding and Incorporates Friction Effects

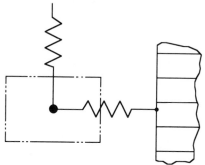

Lumped Mass May Be Applied to One or More DOF at a Node

Figure 7.5 Specialty elements.

Table 7.2 Parameters Affecting Element Behavior and Accuracy

I   Element type
1. Two-dimensional, plane stress/plane strain/axisymmetric
2. Three-dimensional solid
3. Plate
4. Beam

II Integration order
1. Linear
2. Quadratic
3. Linear with added displacement shapes

III Method of formulation
1. Assembly of triangles or tetrahedra
2. Isoparametric

IV Number of integration points
1. Full
2. Reduced
3. Selectively reduced

V   Degree of distortion
1. Aspect ratio
2. Parallelogram angle
3. Trapezoid angle

VI Element Density
1. Number of elements across width
2. Number of elements along length
3. Number of elements across thickness

VII Type of loading
1. Tension
2. Bending, in plane
3. Bending, out of plane
4. Torsion

7.1 ELEMENT INTEGRATION ORDER

For two-dimensional plane stress/plane strain/axisymmetric and three-dimensional solid elements, the feature that subdivides each of these element classes is the element integration order. The

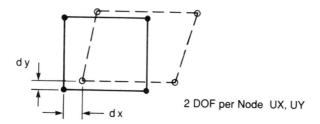

2 DOF per Node UX, UY

First-Order - Constant Strain, Linear Variation in Displacement;
Constant or Average Strain Within Element; Isoparametric

Figure 7.6 Linear displacement element.

basic mathematics of order of integration are explained in Chapter 2, Section 2.9. The order of integration is divided here into three categories: linear displacement (LD) (Figure 7.6), linear displacement with added displacement shapes (LDADS) (Figure 7.7), and quadratic displacement (QD) (Figure 7.8). When inspecting a model, both of the two linear displacement element

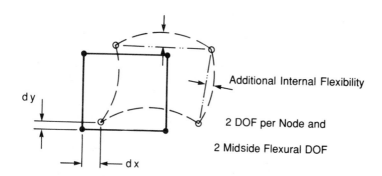

Additional Internal Flexibility

2 DOF per Node and

2 Midside Flexural DOF

First-Order with Additional Displacements
Psuedo Higher-Order
Additional Shapes Allow for Flexing of Element Sides
Subparametric

Figure 7.7 Linear displacement with added displacement shapes element.

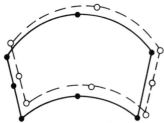

2 DOF per Node : Corner Nodes and
Mid-Side Nodes

Second-Order; Mid – Side Nodes

Quadratic Variation in Displacement
Linear Variation in Strain, Isoparametric
More Efficient, Higher accuracy over First Order
Can Model Curved Boundary Directly
Not as Cost Effective as First-Order with Added Displacement

Figure 7.8 Quadratic displacement element.

categories, LD and LDADS, can be recognized as having nodes
only at their vertices. The quadratic elements can be recognized
by their midside nodes. It is important to note that both of the
linear displacement elements appear identical from the surface;
however, it will be shown that they have significantly different
behavior. This same caveat holds true for other element types
in which elements with identical outward appearance will show
significantly different results, and the user must be careful to
look below the surface to investigate the specific properties of
any element before using or interpreting results.

The linear displacement with added displacement shapes elements almost always show superior performance compared with
the unmodified linear displacement elements. The LDADS elements have two drawbacks, however: 1) They may fail the
"patch test," meaning that the continuity of displacement is
not guaranteed along element boundaries between the nodes as
in the case of the LD elements. 2) There is no equivalent triangular, wedge, or tetrahedron element, so that when these
elements are mixed in a model with LDADS elements, there is
an incompatibility between them, and the triangular, wedge,
and tetrahedra elements are "overstiff." This results in a locally "hard spot" in the model.

The differences between the three element categories are summarized in Table 7.3, along with appropriate figure numbers and references.

7.2 ELEMENT PERFORMANCE TESTS

Due to the number of parameters involved in element performance as outlined in Table 7.2, the only practical way of providing information on the relative and absolute performance of the various element types is by a series of example problems run with various combinations of element types. This sort of test provides an empirical performance standard. A selection of the most commonly used element types are given in Table 7.4, along with their designation to be used in the subsequent test cases.

A set of test cases has been published by MacNeal and Harder [1] as a proposed set of standards to test finite element accuracy. This set of problems represents a good set of benchmark problems in that they begin to approach "real life" problems, but still have exact solutions available for absolute comparison. One difficulty with using very simple shapes such as uniform cylinders, plates, and beams is that they do not allow for the inclusion of the effects of element distortion or geometric discontinuities. The very simplistic comparison problems often show good results with only a few elements, thereby prohibiting any evaluation of element convergence. The following sections give results for the generic element types: two-dimensional plane stress/plane strain/axisymmetric, three-dimensional solid, plate, and beam. These are listed in Table 7.4 for problems taken from the MacNeal—Harder set and others.

The value of a set of benchmark problems is not only that it provides useful information on element performance, but it also provides a good reference against which to test specific finite element codes. One way to get a projection of the accuracy of a particular finite element model would be to select a similar benchmark problem for which an exact solution is available, and to run that problem with the specific finite element code and elements that will be used for the actual model and vary the element density to test its influence on accuracy.

Virtually all commercial finite element codes are undergoing continual development and upgrades, with new releases being issued on a regular basis. Any set of actual element performance numbers, even those generated with an accepted set of benchmark problems such as the MacNeal—Harder set, may be obsolete

Table 7.3 Classes of Element Integration Order

I   Linear displacement: Constant strain
    Linear variation of displacement within element.
    Constant strain (and stress) within element.
    Continuity of displacements across element boundaries, but
      continuity of strain not guaranteed across boundaries.
    Straight-sided elements.
    Applicable to two-dimensional (three-node triangles and four-
      node quadrilaterlas) and three-dimensional elements (six-
      node wedges, eight-node hexahedra, and four-node tetra-
      hedra).
    Ref.: Chapter 2, Section 2.9.
    Figure 6.6.

II Linear displacement with added displacement shapes
    Linear variation of displacement with element midsides al-
      lowed to flex.
    Limited variation in strain across the element.
    Continuity of displacements at the nodes, but compatibility
      of the midside boundaries not guaranteed.
    Straight-sided elements.
    Applicable to two-dimensional (four-node quadrilateral) and
      three-dimensional elements (eight-node hexahedra); ele-
      ments appear to be identical with linear displacement ele-
      ments.
    No triangular, wedge, or tetrahedron elements; must use
      linear displacement triangles and wedges as required.
    Ref.: Chapter 2, Section 1.9.
    Figure 6.7.

III Quadratic displacement: Linear strain
    Quadratic (second-order) variation in displacement within
      element.
    Linear variation of strain (and stress) within element.
    Continuity of displacement and strain across element bound-
      aries.
    Geometry, element sides may be straight or curved (constant
      radius).
    Applicable to two-dimensional elements (six-node triangles,
      eight-node quadrilaterals), three-dimensional elements (10-
      node tetrahedra, 15-node wedges, and 20-node hexahedra),
      and plate (eight-node plates).
    Ref.: Chapter 2, Section 2.9.
    Figure 6.8.

Table 7.4 Designation of Element Types Used in Example Problems

	Integration order		
Geometry	Linear	Linear with added displacement shapes	Quadratic
Two-dimensional			
Triangle	2D-T3		2D-T6
Quadrilateral	2D-Q4	2D-Q4+	2D-Q8
Three-dimensional			
Tetrahedron	3D-T4		3D-T10
Wedge	3D-W6		3D-W15
Hexahedron[a]	3D-H8	3D-H8+	3D-H20
Plate			
Triangle	P-T3		
Quadrilateral	P-Q4		P-Q8
Beam	B-2		

[a]Additional labels of (F), (R), and (S) refer to full $3 \times 3 \times 3$, reduced $2 \times 2 \times 2$, or selectively reduced numerical integration.

in a short space of time. Small changes in element formulation may make no change in the results of certain classes of problems, but it may make significant changes in other classes of problems. Even experienced finite element users may want to develop their own set of benchmark problems, typical of the type of problems they run, and checkout each new release of their finite-element code as it becomes available.

7.3 SIMPLE BEAM IN BENDING

A simple, uniform beam was modeled as a baseline case using each of the elements listed in Table 7.4. The dimensions of the beam, as shown in Figure 7.9, are $1 \times 1 \times 6$, with six elements along the length to give an ideal aspect ratio of 1.0 for the two- and three-dimensional elements. Three load cases were used: tension, in-plane bending, and out-of-plane bending. Because of the square cross section of the beam, only the plate elements

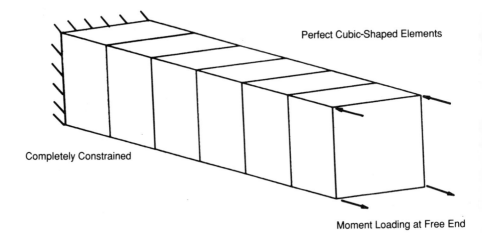

Perfect Cubic-Shaped Elements

Completely Constrained

Moment Loading at Free End

Figure 7.9 Uniform beam 1 × 1 × 6.

used both of the bending load cases. For the other elements, bending in the two directions gives identical results. The intention of this case is to provide a baseline for evaluating the elements under the most simple and ideal conditions possible. Displacement results of the calculations are given in Table 7.5. Results are presented in the form of normalized displacements, i.e., calculated displacements divided by theoretical displacements. Equally important are computer runtimes for the various elements shown in Table 7.6.

These results show that the performance, in bending, of the LDADS elements [2D-Q4+ 3D-H8+(R)] were far superior to that of the LD elements [2D-Q4 3D-H8(R)], both in terms of absolute accuracy and also the ratio of accuracy to runtime. In tension, there is no significant difference in results. This is because the added displacement shapes in the LDADS elements only affect element flexure. In simple tension, the sides of the element remain straight and, therefore, the added displacement shapes have no affect on element performance. Because of the perfect geometry, one would not expect to see any patch-test-type, displacement discontinuity problems.

The LDADS elements showed almost equal accuracy, in bending, to the QD elements (2D-Q8 3D-H20) with a significantly

Table 7.5 Simple Cantilever Beam in Bending 1 × 1 × 6,
Normalized Tip Displacements for Various Element Types

Element type	Number of elements	Tension	Bending Y	Bending Z
2D-Q4+	6	0.996	1.011	
2D-Q4	6	0.995	0.688	
2D-T3	12	0.990	0.242	
2D-Q8	6	1.148	1.005	
3D-H8+(R)	6	0.988	0.994	0.994
3D-H8+(F)	6	0.988	0.994	0.994
3D-H8	6	0.985	0.652	0.652
3D-H20	6	1.250	0.988	0.988
3D-W6	12	0.969	0.218	0.724
P-Q4	6	0.996	1.011	0.975
P-Q8	6	1.277	1.006	1.014
B-2	6	1.277	1.000	1.000

lower cost in computer runtime. For the two-dimensional case, the ratio of LDADS to QD (2D-Q4+ vs. 2D-Q8) was 0.740:1.0 for the total runtime. The ratio of time for individual element stiffness formulation was 0.742:1.0, and the ratio of time for individual element stress calculation was 0.443:1.0. The ratios for the three-dimensional cases were similar with 0.280:1.0 for total runtime, 0.119:1.0 for element formulation, and 0.507:1.0 for stress calculation. Again, due to the simple geometry, this is not a really fair comparison. An advantage of the QD elements is that they can model a curved boundary directly and should be expected to perform relatively better on problems involving curved geometries. For such cases, the QD elements can model a curved boundary with a few elements, whereas the linear elements must use a considerable number of elements to reasonably represent a curved boundary as a series of straight sides.

Table 7.6 Simple Cantilever Beam in Bending 1 × 1 × 6, Runtimes for Various Element Types

Element type	Number of elements	Total time[a]	Element formation time	Stress time
2D-Q4+	6	11.60/2		
2D-Q4	6	10.80/2		
2D-T3	12	10.52/2	0.146	0.061
2D-Q8	6	15.68/2	0.291	0.149
3D-H8+(R)	6	35.92/3	1.442	0.290
3D-H8+(F)	6	70.45/3	4.860	0.972
3D-H8	6	30.85/3	0.747	0.255
3D-H20	6	128.51/3	11.961	0.572
3D-W6	12	52.75/3	1.302	0.279
P-Q4	6	26.32/3	0.616	0.219
P-Q8	6	67.51/3	2.776	0.978
B-2	6	15.55/3	0.045	

[a]Indicates total runtime, in CPU seconds, and number of load cases.

The QD elements showed themselves to be understiff in the tensile loading case for both the two- and three-dimensional elements (2D-Q8, 3D-H20).

The only element to show a difference in bending in the two directions was the wedge element due to the fact that the element pattern is not symmetric. These elements have linear displacements and do not have the added displacement shapes. A comparison of the accuracy of the wedge elements to the 3D-H+ element shows a potential problem when these elements are mixed in the same model. Because there is no LDADS wedge element, the 3D-W6 element used here must be used with the 3D-H+ element to form transition regions when necessary. For this reason, it is recommended that if these two element types must be mixed in a model, the wedge elements should be used sparingly and only in regions of low stress.

A comparison of accuracy vs. runtime for the LD and LDADS elements shows that the superior accuracy of the LDADS elements is achieved without a substantial increase in computer time. The ratios for the two-dimensional LDADS to LD elements were 1.074: 1.0 for total runtime, 1.430:1.0 for element formulation, and 1.0: 1.0 for the stress calculation. Similar ratios were found for the three-dimensional case with 1.164:1.0 for total runtime, 1.930:1.0 for element formulation, and 1.137:1.0 for stress calculation.

As expected, the beam element behaved very economically, giving very high accuracy with the lowest runtime. This is due, in large part, to the rotational degrees of freedom that transmit moments directly between nodes. The plate elements also did well, considering that the ratio of thickness to length and width of 1.0 is not recommended. The computer runtime for the plate elements, obviously, is greater than that of the beam elements due to the fact that twice the number of nodes are involved.

A second simple beam is shown in Figure 7.10 to check the rate of convergence of three simple two-dimensional elements: 2D-Q4, 2D-T3, and 2D-T6. This case modeled a beam of a rectangular cross section with dimensions 1 × 5 × 10 long with an in-plane bending load. This case used maximum bending stress as its criteria for evaluation of accuracy. Results are presented as normalized stresses, i.e., maximum calculated centroidal bending stresses divided by theoretical bending stresses, $\sigma = Mc/I$, at the same location. This beam case illustrated some fundamental principles of element convergence and general behavior.

1. The rates of convergence (Figure 7.11 and Table 7.7) showed that the six-node triangle (2D-T6) converged the fastest, followed by the four-node quadrilateral without added displacement shapes (2D-Q4) and the three-node triangle (2D-T3), when the convergence criterion was accuracy vs. number of elements.

2. When the convergence criterion was accuracy vs. number of degrees of freedom in the model, Figure 7.12 shows that the number and type of element were not significant in this case. This should not be taken as a blanket conclusion covering all element types, structures, and loading cases. When the LDADS elements are used, they have the same number of degrees of freedom as the LD elements and dramatically different results.

RESULTS of CANTILEVER BEAM
VARIABLE MODEL

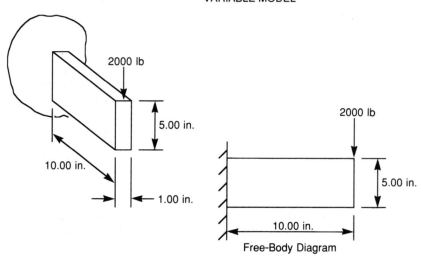

Free-Body Diagram

Model with Two-Dimensional
Linear Quadilateral Slab Elements

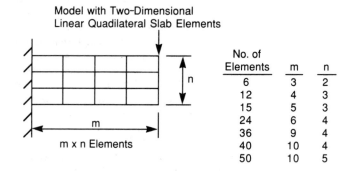

No. of Elements	m	n
6	3	2
12	4	3
15	5	3
24	6	4
36	9	4
40	10	4
50	10	5

m x n Elements

Figure 7.10 Simple beam used for convergence tests.

3. The total computer runtime is a strong function of the number of degrees of freedom (Figure 7.13, Table 7.8, and Table 7.9) for these element types. The number of elements has only a secondary effect although the number of degrees of freedom and number of elements are closely related.

Model with 2D Linear and Quadratic Triangular Elements

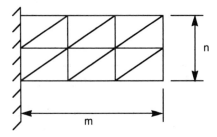

m x n Elements

n

m

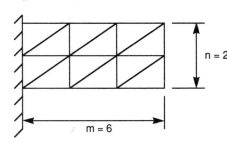

n = 2

Example
6 x 2 = 12
Elements

m = 6

No of Elements	m	n
12	6	2
18	6	3
24	8	3
30	10	3
48	12	4

7.4 CURVED-BEAM BENCHMARK CASE

This case was taken from the MacNeal–Harder paper [1]. Dimensions of the beam are shown in Figure 7.14 and are inner radius = 4.12, outer radius = 4.32, arc = 90°, thickness = 0.10, Young's modulus = 10,000,000, and Poisson's ratio = 0.25. Unit loading at the beam tip is in both in-plane and out-of-plane directions.

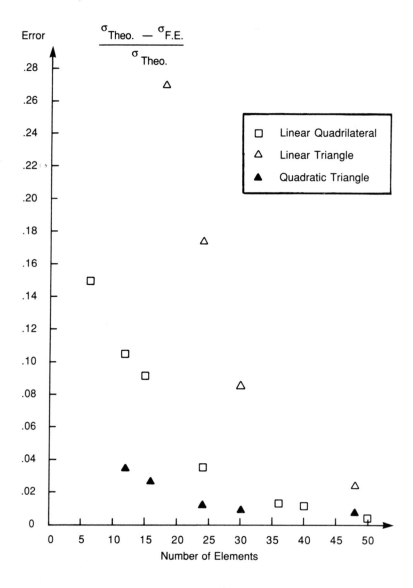

Figure 7.11 Accuracy vs. number of elements for three two-dimensional elements.

Table 7.7 Simple Beam: Variable Mesh Density,
Normalized Stress Results

Mesh density	2D-Q4	2D-T3	2D-T6
10 × 5	0.997	0.921	
10 × 4	1.012		
9 × 4	1.015		
6 × 4	1.039	1.023	0.994
5 × 3	1.066	1.022	1.000
4 × 3	1.096	1.096	1.013
3 × 3	1.153	1.271	1.017
3 × 2	1.150		1.037

The exact solution at the beam tip is as follows:

In-plane displacement (due to unit loading) = 0.08734

Out-of-plane displacements (due to unit loading) = 0.5022

Results are, once again, presented as normalized displacements.

Tests were run with eight different element types: two two-dimensional elements (LD and LDADS, Figure 7.15), three three-dimensional elements (LD, LDADS, and QD, Figure 7.16), two plate elements (linear and quadratic), and a beam element (Figure 7.17. The results of all runs are tabulated in Table 7.10. For purposes of comparison of the effects of element type with the same mesh density, the case of six elements along the arc by one element across the section should be noted.

As expected, the LDADS elements outperformed the linear elements for both the two- and three-dimensional cases. For the three-dimensional cases, the QD solid element did not perform better than the LDADS elements. The only significant difference was for the out-of-plane loading cases in which the QD element (3D-H20) performed better than the LDADS element [3D-H8+(R)]. The QD plate element (P-Q8) showed better performance than its LD counterpart (P-Q4), which is to be expected because the QD plate can match exactly the curvature of the beam. It is somewhat surprising that the QD solid element did not perform relatively better because it matched exactly the curved boundary,

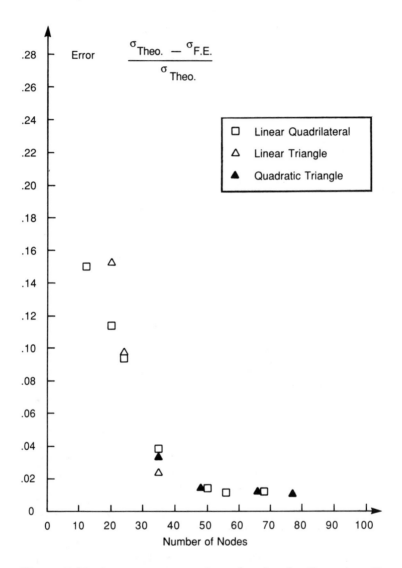

Figure 7.12 Accuracy vs. number of nodes for three two-dimensional elements.

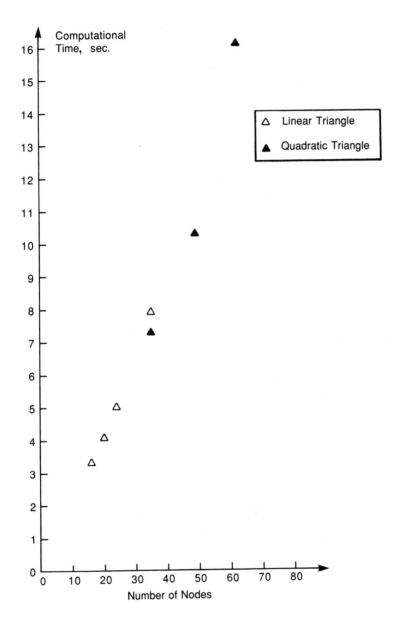

Figure 7.13 Computer time vs. number of nodes for three two-dimensional elements.

Table 7.8 Simple Beam: Variable Mesh Density,
Number of Degrees of Freedom

Mesh density	2D-Q4	2D-T3	2D-T6
10 × 5	120	132	
10 × 4	110	70	234
9 × 4	100		
6 × 4	70	48	234
5 × 3	48	48	154
4 × 3	40	40	126
3 × 3	32	32	98
3 × 2	24	24	70

whereas the other elements [2D-Q4, 2D-Q4+, 3D-H8(R), 3D-H8+(R), P-Q4] modeled the curved boundary as a series of straight segments.

The variable element density study using the QD, solid element (3D-H20) shows that although a single element can match the geometry of the beam, the one- and three-element model did not give adequate results. This is due to the extreme (90°) arc of the model.

Table 7.9 Simple Beam: Variable Mesh Density, Number of Degrees of Freedom and Computer Runtime

Mesh density	2D-T3 DOF/sec	2D-T6 DOF/sec
10 × 4	70/7.6	234/30.1
5 × 3	48/5.0	154/22.5
4 × 3	40/4.3	126/16.2
3 × 3	32/3.2	98/10.5
3 × 2	24/2.6	70/7.2

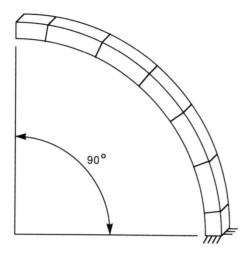

May Be Modeled with Plate/Beam/2D/3D Solid Elements
90 degree Arc   Outer Radius = 4.32   Inner Radius = 4.12
Thickness = 0.1   Young's Modulus = 1,000,000   Poisson's Ratio = 0.3
Unit Loading Applied at Free End - In Plane and out of Plane

Figure 7.14 Curved-beam benchmark case.

7.5 TWISTED-BEAM BENCHMARK CASE

This case was also taken from the MacNeal—Harder paper [1]. Dimensions of the beam are shown in Figure 7.18. The dimensions are length = 12.0, width = 1.1, depth = 0.32. There is a 90° twist along the length. Young's modulus = 29,000,000, Poisson's ratio = 0.22, and there is unit loading at the tip in both in-plane and out-of-plane directions.

The exact solution is

In-plane displacement (unit loading) = 0.005424

Out-of-plane displacement (unit loading) = 0.001754

Results are presented as normalized displacements.

Tests were run with six different element types with a pattern of 12 elements along the length and two elements across the width. The results are given in Table 7.11. In addition, the element

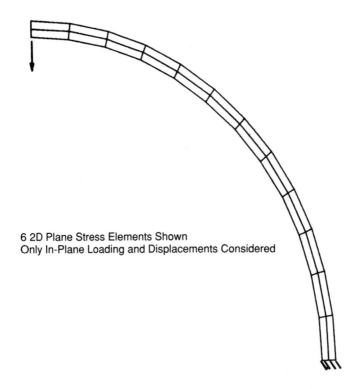

6 2D Plane Stress Elements Shown
Only In-Plane Loading and Displacements Considered

Figure 7.15 Curved-beam two-dimensional model.

density was varied for the 3D-H8+(R) and P-T3 elements (Figures 7.19 and 7.20) and the results given in Table 7.12.

The results show that, for the 12 × 2 element pattern, all element types with the exception of the LD solid [3D-H8(R)] gave
good results. The LD solid element was, again, overstiff and
gave unacceptable results. This twisted-beam case allows for
a study of the effects of distortion (over 8° of warping per element). A comparison of the results of this case with the results
of the baseline, square beam given in Table 7.5 shows comparable results with no significant degradation in accuracy due to
the warping in the twisted-beam case, with the exception of the
LD solid [3D-H8(R)] element. The elements that gave acceptable
results for the square beam case still gave acceptable results for
the twisted-beam case.

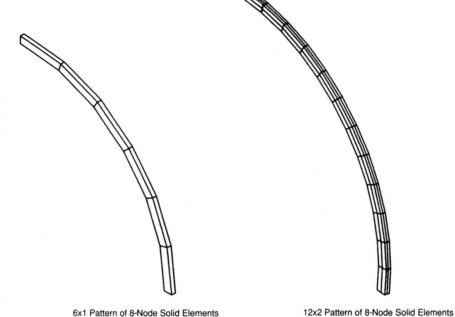

6x1 Pattern of 8-Node Solid Elements   12x2 Pattern of 8-Node Solid Elements

Figure 7.16 Curved-beam three-dimensional models.

   In the case of the plate elements, both the quadrilateral and
triangular elements were run. The triangular elements, by definition, do not have any warping. Table 7.11 shows that for the
12 × 2 gridwork for quadrilaterals (12 × 4 for triangles), there
was no significant advantage to the triangular elements despite
the fact that there were twice as many triangles as quadrilaterals
used in the models. In fact, the triangular element showed an
almost 20% error in the case of in-plane loading.
   Table 7.12 gives the results for variable element density for
two element types, the P-T3 plate and the 3D-H8+(R) solid.
These results show that for the in-plane loading condition, there
was little variation in accuracy for the plate element and only a
significant difference in results for the solid element when the
mesh was reduced to three elements. These results show that
both elements approach a point of being about 20% understiff for
the solid element and 10% for the solid element.

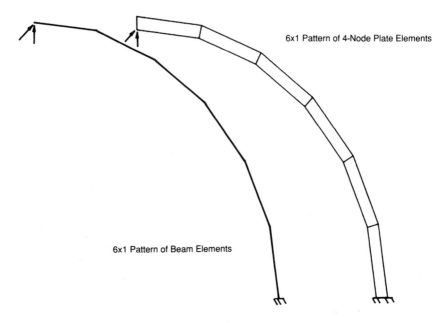

6x1 Pattern of 4-Node Plate Elements

6x1 Pattern of Beam Elements

Figure 7.17 Curved-beam plate and beam models.

The solid element showed that with more elements, the overall stiffness of the model decreases, which is generally the case for most elements. This variation in normalized displacement from 0.785–1.096 for the in-plane loading may give the misleading impression that there is an optimum number of elements for this model, beyond which an increase in the number of elements causes a decrease in accuracy. For this particular case, there is an apparent optimum mesh of 6 × 1 × 1, which gives a normalized displacement of 1.001, i.e., 0.1% from the theoretical solution. It is important to recognize that this crossover point is a combination of two phenomena: (1) This particular LDADS element converges to high displacement results for this model under this loading condition, and (2) the decrease in flexibility as a function of the decreasing number of elements.

For the out-of-plane loading condition, the plate element was slightly understiff in the limit, but overstiff for the coarsest model. For the solid element, the trend was, once again, increased flexibility as a function of the number of elements. In

Table 7.10 Results of Curved-Beam Test Cases

Element type	Mesh density	Normalized displacement	
		In plane	Out of plane
2D-Q4	6 × 1	0.244	
	6 × 2	0.737	
	12 × 1	0.244	
	12 × 2	0.247	
2D-Q4+	3 × 1	0.186	
	6 × 1	0.824	
	6 × 2	0.822	
	12 × 1	0.998	
	12 × 2	0.996	
3D-H8	6 × 1	0.073	
3D-H8+(R)	3 × 1	0.439	
	6 × 1	0.935	
	12 × 1	1.004	
3D-H8+(S)	6 × 1	0.990	0.849
3D-H20	1 × 1	0.009	
	3 × 1	0.309	
	6 × 1	0.875	0.946
P-Q4	3 × 2	0.186	
	6 × 1	0.833	0.951
	6 × 2	0.822	
	12 × 1	0.998	
	12 × 2	0.996	
P-Q8	6 × 1	1.007	0.971
B-2	6 × 1	0.904	0.973

this loading condition, however, the element was still somewhat overstiff in the limit. In contrast to the in plane loading condition that was 9.6 over the theoretical displacement, the out-of-plane loading condition gave results 11.4% below the theoretical displacement. There was a fairly uniform convergence of displacement as a function of element density from 0.573--0.886.

May Be Modeled with Solid or Plate Elements
Length = 12.0   Width = 1.1   Depth = 0.32
90 degree Twist
Young's Modulus = 30,000,000   Poisson's Ratio = 0.3
Unit Loading at Free End - in Plane and out of Plane

Figure 7.18 Twisted-beam benchmark case.

7.6 RECTANGULAR PLATE BENCHMARK CASE

This case is taken from the MacNeal—Harder paper [1]. This model is of two simple rectangular plates, each with two sets of boundary constraints: simply supported and clamped. The plates are shown in Figure 7.21. Their dimensions are 2 × 2 and 2 × 10 to give two different geometries. The thickness is 0.0001 for the models using plate elements and 0.01 for the solid-element models. Young's modulus is 17,420,000 and Poisson's ratio 0.3. Two loading conditions are applied, a uniform loading of q = 0.0001 force/area and a concentrated loading, at the centroid of 0.0004 force.

Table 7.11 Results of Twisted-Beam Test Cases: Six Element Types, 12 × 2 Element Grid Pattern

	Normalized displacements					
	P-Q4	P-T3	P-Q8	3D-H8	3D-H8+(S)	3D-H20
In plane	0.993	1.198	0.998	0.336	0.938	0.991
Out of plane	0.985	0.998	0.998	0.202	0.977	0.995

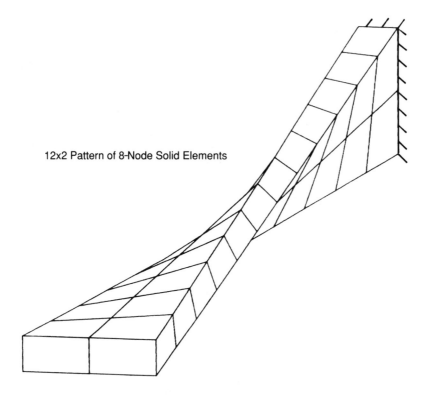

12x2 Pattern of 8-Node Solid Elements

Figure 7.19 Twisted-beam three-dimensional solid model.

Models are formed with four different element types; two plate
elements, P-Q4 and P-Q8; and two solid elements, 3D-H8+(R) and
3D-H20. For each of the element types, four element densities
are used: 2 × 2, 4 × 4, 6 × 6, and 8 × 8. The model is quarter-
symmetric so that the element densities refer to the number of the
elements in the quarter of the plate.

The theoretical solution is taken from Timoshenko [2] and is
given in Table 7.13.

Results in terms of normalized displacements at the center of
the plate are given in Table 7.14 for the 2 × 2 plate and in Table
7.15 for the 2 × 10 plate. The results show good accuracy with
both the plate elements for both geometries for the uniform loading, simply supported cases. For the concentrated load with

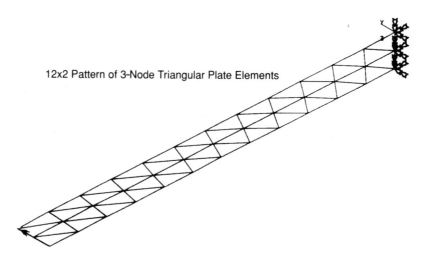

12x2 Pattern of 3-Node Triangular Plate Elements

Figure 7.20 Twisted-beam plate model.

Table 7.12 Results of Twisted-Beam Test Cases: Two Element Types, Variable Mesh Density

	Mesh density	P-T3	3D-H8+(R)	3D-H8+(S)
In plane	12 × 4 × 1	1.198		
	12 × 2 × 1		1.096	0.983
	12 × 1 × 1		1.077	
	6 × 2 × 1	1.189	1.062	
	6 × 1 × 1		1.001	
	3 × 2 × 1	1.180	0.945	
	3 × 1 × 1		0.785	
Out of plane	12 × 4 × 1	1.042		
	12 × 2 × 1	0.998	0.886	0.977
	12 × 1 × 1		0.871	
	6 × 2 × 1	0.930	0.838	
	6 × 1 × 1		0.792	
	3 × 2 × 1	0.798	0.685	
	3 × 1 × 1		0.573	

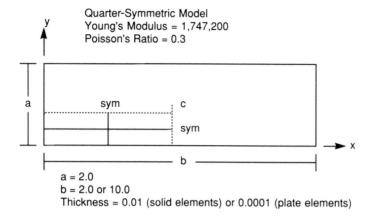

Figure 7.21 Rectangular plate benchmark case.

Table 7.13 Theoretical Displacements for a Rectangular Plate

	Displacement at center of plate	
	Uniform loading	Concentrated load
2 × 2 Plate		
Simply supported	4.062[a]	11.60
Clamped	1.26	5.60
2 × 10 Plate		
Simply supported	12.97	16.96
Clamped	2.56	7.23

[a]4.062 is the theoretical displacement for the plate-element model; 4.062×10^{-6} is the displacement for the solid-element model. The thickness of the solid-element model is 100 times greater than that of the plate-element model.
Source: Ref. 2.

Table 7.14 Results of 2 × 2 Rectangular Plate, Normalized Displacements at the Center of the Plate

Element type	Mesh density	Uniform load, simply supported	Concentrated load, clamped supports
P-Q4	2 × 2	0.981	0.934
	4 × 4	1.004	1.010
	6 × 6	1.003	1.012
	8 × 8	1.002	1.010
P-Q8	2 × 2	0.927	1.076
	4 × 4	0.996	0.969
	6 × 6	0.999	0.992
	8 × 8	1.000	0.997
3D-H8+(S)	2 × 2	0.989	0.885
	4 × 4	0.998	0.972
	6 × 6	0.999	0.988
	8 × 8	1.000	0.994
3D-H20	2 × 2	0.023	0.002
	4 × 4	0.738	0.072
	6 × 6	0.967	0.552
	8 × 8	0.991	0.821

clamped supports, the plate elements gave good results for the 6 × 6 and 8 × 8 element densities for the 2 × 10 plate model and good results for all element densities for the 2 × 2 plate model. For the coarser meshes, 2 × 2 and 4 × 4, there is an apparent effect of the length-to-width aspect ratio. For these cases, there is no apparent advantage to using the quadratic plate element (P-Q8) over the linear plate (P-Q4).

For the models constructed with solid elements, length-to-width aspect ratio has an effect on the LDADS [3D-H8+(R)] element, especially in the case of the concentrated loading with clamped supports. The QD solid element (3D-H20) showed a pronounced effect of both length-to-width and length (or width-)-to-thickness aspect ratios. Although the model constructed with solid elements is 100 times thicker than the plate-element models,

Table 7.15 Results of 2 × 10 Rectangular Plate, Normalized Displacements at the Center of the Plate

Element type	Mesh density	Uniform load, simply supported	Concentrated load, clamped supports
P-Q4	2 × 2	1.052	0.519
	4 × 4	0.991	0.863
	6 × 6	0.997	0.940
	8 × 8	0.998	0.972
P-Q8	2 × 2	1.223	0.542
	4 × 4	1.003	0.754
	6 × 6	1.000	0.932
	8 × 8	1.000	0.975
3D-H8+(S)	2 × 2	0.955	0.321
	4 × 4	0.978	0.850
	6 × 6	0.990	0.927
	8 × 8	0.995	0.957
3D-H20	2 × 2	0.028	0.001
	4 × 4	0.693	0.041
	6 × 6	1.066	0.220
	8 × 8	1.026	0.374

the thickness of the solid elements is still only 0.01. For the 2 × 2 plate model with an 8 × 8 mesh, each element is 0.125 × 0.125, which gives the best-case length-to-thickness aspect ratio of 12.5. The worst-case length-to-thickness aspect ratio is 100.0 for the 2 × 10 plate with the 2 × 2 mesh. Figure 7.22 shows the variation in normalized displacement as a function of aspect ratio.

In a sense, this is an unfair comparison of plate elements vs. solid elements. If the computer runtimes are also taken into account, the plate elements would appear to be far superior. In comparing the results of the plate element models with the Timoshenko closed-form solution, there should be good agreement. The same basic force/deflection relationships in the closed-form solution are also present in the plate element formulation. There are many cases, in practice, in which a plate-type component must

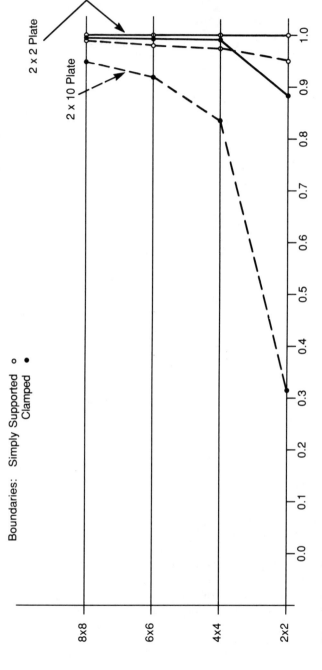

Loading: Uniform Pressure = 0.0001 or Central Point Load = 0.0004

Boundaries: Simply Supported ○
Clamped ●

2 x 2 Plate

2 x 10 Plate

Figure 7.22 Normalized displacement vs. aspect ratio: Rectangular plate benchmark case, eight-node solid elements.

be modeled with three-dimensional solid elements as part of a larger model with fully three-dimensional geometry details that require the use of three-dimensional solid elements. This case illustrates some of the capabilities and shortcomings of the use of solid elements in modeling flat plates.

It is interesting to note the effect of loading and constraint conditions on model accuracy. A comparison of the two columns in either Table 7.15 or 7.16 shows these effects. In most cases, the models tended to underpredict the theoretical displacements. However, in a few cases (P-Q8 element), the models overpredicted the displacement for one loading/constraint case and underpredicted the displacement for the other case with the same element density and geometry. When these effects are taken into account with the aspect ratio effects, it becomes clear that the performance of a particular element cannot be completely deduced from an isolated test.

7.7 SCORDELIS—LO ROOF BENCHMARK CASE

This case was first published by Scordelis and Lo in 1969 [3], discussed by Zienkiewicz in his text [4], later included in the MacNeal and Harder proposed benchmark set [1], and is becoming a classic among finite element benchmark cases.

The structure and model are shown in Figure 7.23. It is a curved-plate-type structure, length = 50, thickness = 0.25, and the "width" specified as a 25.0 radius with an 80° included arc. The Young's modulus is 432,000,000 and Poisson's ratio 0.0. The loading is -90.0 in the vertical (Z) direction. One-quarter of the structure is included in the model with symmetry boundary conditions applied on the cut boundaries. Axial and radial displacements are constrained to 0 on the curved boundaries.

Calculations are run using four element types; two plate elements, P-Q4 and P-Q8; and two solid elements, 3D-H8+(R) and 3D-H20. The results are presented in Table 7.16.

The length-to-thickness aspect ratio varies between 20 and 100, which is about the same range as the models for the flat rectangular plate. It comes as no surprise that the results for the solid-element models in the coarsest mesh are poor. The results for the QD solid element (3D-H20) are similar to the results for the flat rectangular plate case. It is interesting and somewhat disappointing to note that the QD solid element underperformed the LDADS element for all meshes. With the additional

Boundaries: Radial and Axial Constraints on Curved Free Boundaries
             Symmetry on Cut Boundaries

Radius = 25.0
Thickness = 0.25
Length = 50.0

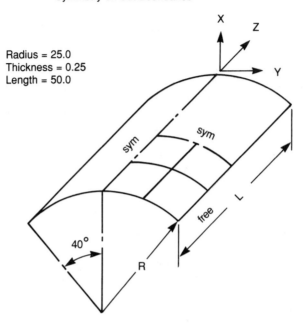

Young's Modulus = 432,000,000
Poisson's Ratio = 0.0
Distributing Loading = 90.0 per unit area in -z direction

Figure 7.23 Scordelis–Lo benchmark case: Concrete cylindrical shell quarter-symmetric model.

midside node, the QD solid is able to model exactly a cylindrical surface. The LDADS element must approximate the surface as a series of flats. The logical explanation is that the length-to-thickness aspect ratio affects the QD solid more adversely than the LDADS solid element.

Both plate elements performed well with virtually identical results at the more refined meshes. The QD plate element (P-Q8) converges to within 0.2% of the theoretical value with the 6 × 6 mesh and, therefore, cannot show any more improvement with

Table 7.16 Results of Scordelis–Lo Roof, Normalized
Displacements at the Center of the Free Edge

Element type	Mesh density	Uniform load, simply supported
P-Q4	2 × 2	1.376
	4 × 4	1.050
	6 × 6	1.018
	8 × 8	1.008
	10 × 10	1.004
P-Q8	2 × 2	1.021
	4 × 4	0.984
	6 × 6	1.002
	8 × 8	0.997
	10 × 10	0.996
3D-H8+(S)	2 × 2	1.320
	4 × 4	1.028
	6 × 6	1.012
	8 × 8	1.005
3D-H20	2 × 2	0.092
	4 × 4	0.258
	6 × 6	0.589
	8 × 8	0.812

more refined meshes. With the coarser meshes, there can be
problems with the QD plate elements, such as the P-Q8 when the
out-of-plane curvature exceeds a certain limit. A rule-of-thumb
limit is that the height of the arc for each element should not be
greater than 10% of the chord. In this case, the ratio is 0.088
for the 2 × 2 mesh (i.e., each element including a 20° arc) and
0.017 for the 10 × 10 case. This 10% limit number corresponds
to a 22.5° arc.

7.8 SPHERICAL SHELL BENCHMARK CASE

This spherical shell benchmark case gives a model with double
curvature. The theoretical solution is given by Morely and Morris

[5] for a complete spherical shell. The case as it is presented here was modified by MacNeal and Harder to include a hole at the top of the sphere to avoid the need for triangular elements [1].

The structure and model are shown in Figure 7.24. The dimensions of the hemisphere are radius = 10.00 and thickness = 0.04 with a hole at the top subtended by an 18° angle. Young's modulus is 6.825×10^7 and Poisson's ratio is 0.3. Forces are applied at four points, 90° apart, along the free edge of the hemisphere in the radial direction, as shown. The forces have a magnitude of 2.0, but alternate in sign. One-quarter of the hemisphere is modeled with symmetry boundary conditions along the cut boundaries. Models are constructed from four element types; two plate elements, LD (P-Q4) and QD (P-Q8); and two solid elements, LDADS [3D-H8+(R)] and QD (93D-H20). Five element densities are used for the plate elements and three element densities for the solid elements.

The theoretical solution given by Morley and Morris for displacement at the point of force application is 0.0924 for a complete hemisphere. MacNeal and Harder used a value of 0.0940 for the hemisphere with a hole at the top.

Results are presented in Table 7.17. A comparison of the two plate elements shows that the LD element (P-Q4) showed good behavior for all element densities. The QD element showed poor performance for the coarse element densities with a uniform convergence at the finer element densities and gave similar results for the finest, 12 × 12 mesh. The behavior of the QD plate is substantially worse in this case than in the cases of the rectangular plate and the Scordelis–Lo roof. The explanation is the double curvature of the elements. For the coarsest, 2 × 2 mesh, each element must span 45° along the equator (free surface) and 36° along the symmetric boundaries.

The solid elements showed a similar pattern with the LDADS element [3D-H8+(R)] giving better performance than the QD element (3D-H20). The LDADS element gave a good result for the 12 × 12 mesh. The QD element gave unacceptable results for all meshes.

The same comments regarding the selection of plate elements vs. solid elements that were made for the rectangular plate also apply to the Scordelis–Lo roof and the spherical shell. In modeling plate-type structures, plate elements will perform at least as well as solid elements. When the computer runtime is also taken into account, plate elements become the clear winner. The choice to use solid elements will be necessitated by the need to model details such as lugs, fitting, ports, gussets, ribs, etc. and

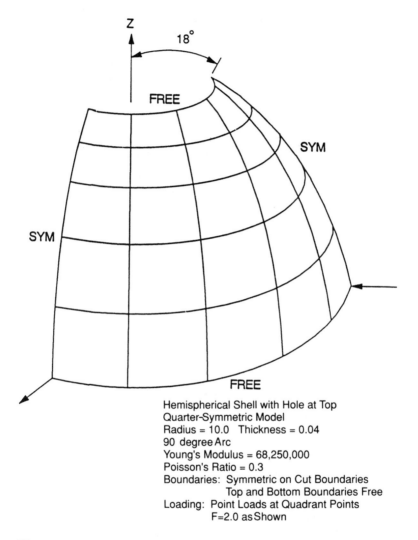

Hemispherical Shell with Hole at Top
Quarter-Symmetric Model
Radius = 10.0   Thickness = 0.04
90 degree Arc
Young's Modulus = 68,250,000
Poisson's Ratio = 0.3
Boundaries: Symmetric on Cut Boundaries
            Top and Bottom Boundaries Free
Loading: Point Loads at Quadrant Points
         F=2.0 as Shown

Figure 7.24 Spherical shell benchmark case.

obtain accurate and detailed stresses. In such cases, an alternative to solid elements is the use of plate elements throughout with
a second, refined model of the geometric details. Refined models
are discussed in more detail in Chapter 9.

Table 7.17 Results of Spherical Shell Test Case, Normalized Radial Displacement at the Point of Load Application

Element type	Mesh density	Uniform load, simply supported
P-Q4	2 × 2	0.972
	4 × 4	1.024
	6 × 6	1.013
	8 × 8	1.005
	10 × 10	1.001
	12 × 12	0.998
P-Q8	2 × 2	0.025
	4 × 4	0.121
	6 × 6	0.494
	8 × 8	0.823
	10 × 10	0.955
	12 × 12	0.955
3D-H8+(S)	4 × 4	0.039
	8 × 8	0.730
	12 × 12	0.955
3D-H20	4 × 4	0.001
	8 × 8	0.776
	12 × 12	0.972

7.9 THICK-WALLED CYLINDER

This geometrically simple problem is intended to test the ability of the various elements to model a nearly incompressible material with a Poisson's ratio approaching 0.5. This problem is taken from the MacNeal—Harder set of test cases [1]. The structure and model is shown in Figure 7.25. The cylinder has an inner radius of 3.00, an outer radius of 9.00, and a length of 1.00. There is a unit pressure acting on the inner surface. Young's modulus is set to 1000 and Poisson's ratio is varied as 0.49, 0.499, and 0.4999. A 5 × 1 gridwork representing a 10° segment of the cylinder is used for the model.

A theoretical solution is given by MacNeal and Harder [1] and Roark [6]

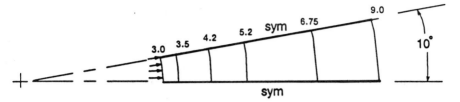

Plane Strain Model for Nearly Incompressible Material
Inner Radius = 3.0 Outer Radius = 9.0 Thickness = 1.0
Young's Modulus = 1000.
Poisson's Ratio = 0.49 0.499 0.4999
Loading: Unit Pressure on Inner Surface

Figure 7.25 Thick-walled cylinder benchmark case.

$$u = \frac{(1+\mu)\ p\ R^2_{inner}}{E(R^2_{outer} - R^2_{inner})}\ [R^2_{outer}/r + (1-2\mu)r] \qquad (7.1)$$

For this model, the solutions for three values of Poisson's ratio are

Poisson's ratio	Radial displacement at inner radius
0.49	0.00050399
0.499	0.00050602
0.4999	0.00050623

Eight different element types are used with the three values of Poisson's ratio as shown in Table 7.18. The same 5 × 1 gridwork is used for all elements. Boundary conditions are specified to represent symmetry along the cut boundaries by permitting displacement in the radial direction and constraining against displacement in the tangential direction. Plane strain conditions are assumed for the two-dimensional and plate elements.

Table 7.18 shows that there is no substantial difference between the LD, LDADS, and QD two-dimensional elements, with the LD and QD elements slightly overpredicting displacement by the same amount for all cases of Poisson's ratio and the LDADS element underpredicting slightly. This is to be expected because the added displacement shapes only affect the element's

Table 7.18 Results of Thick-Walled Cylinder Test Case,
Normalized Radial Displacement at Inner Boundary

| | Variable Poisson's ratio | | |
| | Poisson's ratio | | |
Element type	0.49	0.499	0.4999
2D-Q4	1.019	1.020	1.020
2D-Q4+	0.988	0.982	0.931
2D-Q8	1.036	1.037	1.037
P-Q4	0.846	0.359	0.053
P-Q8	1.000	0.997	0.967
3D-H8	0.936	0.809	0.771
3D-H8+(S)	0.986	0.986	0.986
3D-H9+(R)	1.025	1.026	1.026
3D-H20	0.999	0.986	0.879

performance in bending. In cases of pure tensile or compressive
loading, there should be no difference in behavior.

For the three-dimensional solid elements, the LDADS element
[3D-H8+(R)] showed uniform behavior, whereas the QD element
(3D-H20) showed a decrease in accuracy for the case of Poisson's
ratio equal to 0.4999 and the LD element [3D-H8(R)] showed a
decrease in accuracy across all cases of Poisson's ratio. The LD
plate element (P-Q4) had difficulty with the cases of Poisson's
ratio equal to both 0.499 and 0.4999.

7.10 ROTATING DISK BENCHMARK CASE

This case of a hyperbolic disk rotating about its axis, which is
not part of the MacNeal—Harder set, gives a case of nonuniform
body (centrifugal) forces and an opportunity to investigate the

mass properties of the various element types. This type of problem has practical interest in turbomachinery applications, which provide the basis for the theoretical solution [7]. Timoshenko [8] offers a closed-form solution and discussion.

The disk in this benchmark case is described by Roark [6] and has a hyperbolic profile described as

$$t = cr^a \qquad (7.2)$$

where t is the disk thickness at radius, r and c and a are constants. The disk used in this benchmark case is shown in Figure 7.26, where c = 4.819 and 1 - −0.6826 so that t - 3.00 in. at r_1 (inner radius) and t = 1.00 in. at r_2. The Young's modulus is 30,000,000 psi, Poisson's ratio 0.3, and the weight density 0.279 lbf/in.^{3}. The rotational speed is 377 rad/sec.

The theoretical solution for stress is taken from Roark [6] and is based on the solution by Stodola [7]. Roark's notation is used.

$$\sigma_r = \frac{E}{1 - \mu^2}[(3+\mu)Fr^2 + (m_1+\mu)Ar^m1^{-1} + (m_2+\mu)Br^m2^{-1}] \text{ psi} \qquad (7.3)$$

$$\sigma_t = \frac{E}{1 - \mu^2}[(1+3\mu)Fr^2 + (1+m_1\mu)Ar^m1^{-1} + (1+m_2\mu)Br^m2^{-1}] \text{ psi} \qquad (7.4)$$

where

σ_r and σ_t = radial and tangential (hoop) stresses

E = Young's modulus, 30,000,000 psi

μ = Poisson's ratio, 0.3

r = the distance from the centerline

$$F = \frac{-(1-\mu^2)\,\delta\,w^2\,/\,386.4}{E[8 + (3+\mu)\,a]} \qquad (7.5)$$

δ = the weight density, 0.279 lbf/in.^{3}

w = the rotational speed, 377.0 rad/sec

a = the exponent from the hyperbolic thickness expression, as shown above, -0.6826

$$m_1 = -\frac{a}{2} - \left(\frac{a^2}{4} - a\mu + 1\right)^{1/2} \qquad (7.6)$$

$$m_2 = -\frac{a}{2} + \left(\frac{a^2}{4} - a\mu + 1\right)^{1/2} \qquad (7.7)$$

A and B are constants derived by using known radial stresses at the boundaries and solving Equation 7.5 as a set of two simultaneous equations with r = r_1 and r = r_2 together with the appropriate radial stresses. In this case, radial stresses at the bore

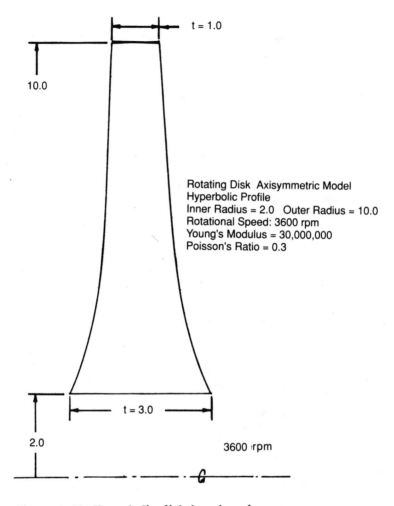

Figure 7.26 Hyperbolic disk benchmark case.

bore and at the outside diameter, set equal to zero, are used as boundary conditions.

$$r = r_1 = 2.00, \qquad \sigma_{r1} = 0.0$$
$$r = r_2 = 10.00, \qquad \sigma_{r2} = 0.0$$

The following values of A and B are obtained.

A = 5.30251×10^{-4}

B = 3.34593×10^{-4}

These constants satisfy the boundary conditions of $\sigma_r = 0.0$ at r = 2.00 and 10.00. Radial and tangential stresses are calculated at two other points that correspond to element centroid locations for two gridwork patterns used in the test case.

r = 2.4 (centroid for 20-element gridwork)
σ_r = 866.99 psi σ_t = 4974.46 psi

r = 2.8 (centroid for 5-element gridwork)
σ_r = 1424.97 psi σ_t = 4434.14 psi

It should be noted that the solution is extremely sensitive to the constants A and B that must be precisely calculated to at least five significant figures. A simple calculation with a hand calculator may not be adequate to obtain A and B with sufficient precision. It is recommended that in order to use Equations 7.3– 7.7 to calculate other comparison cases, a computer should be used to achieve the required precision.

Three axisymmetric element types are used in the models: LD (2D-Q4), LDADS (2D-Q4+), and QD (2D-Q8). All three element types are used with a 5 × 1 gridwork as shown in Figure 7.27. The LD and LDADS elements are also used with a 2 × 10 gridwork. Stresses are calculated at the element centroids for comparison with the theoretical solution. A limiting case of a single element was also run. In the single element model, however, the disk cross section is a simple tapered section and not a hyperbola.

The results of normalized centroidal stresses, i.e., calculated FEA stress divided by the theoretical stress at the same radius, are presented in Table 7.19. These results show that calculation of tangential (hoop) stress is consistently more accurate than the calculation of radial stress. The tangential stress results are good with both the 5-element and 20-element models for all element types. For radial stress, the LD element (2D-Q4) gave the best results, nominally 8% low, with the LDADS element (2D-Q4+) giving results about 13–19% low. The quadratic element underpredicted stress by about 12%. It may appear somewhat surprising that the QD element did not perform better, considering that it should approximate the hyperbolic surface with a

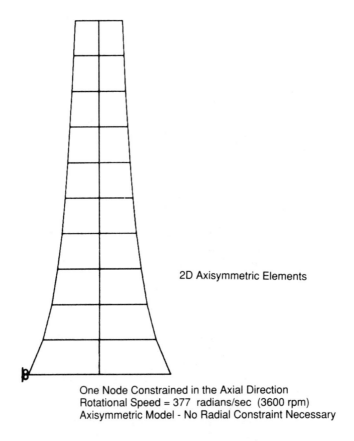

2D Axisymmetric Elements

One Node Constrained in the Axial Direction
Rotational Speed = 377 radians/sec (3600 rpm)
Axisymmetric Model - No Radial Constraint Necessary

Figure 7.27 Hyperbolic disk axisymmetric model.

second-order curve. A comparison of total mass moment of each model is given in Table 7.20 and shows that despite the fact that the sides of the QD element can model a curved surface, the mass moment numbers show no difference between the two, 5-element linear and LDADS models and the QD model. The two, 20-node models showed slightly lower mass moments as expected because the elements will always overpredict volume (and mass) on a concave surface.

Table 7.19 Results of Rotating Hyperbolic Disk, Normalized Radial and Hoop Stress (rotational speed = 377 rad/sec)

Element type	Number of elements	Radial location	Radial stress	Hoop stress	Total mass
2D-Q4	1[a]	2.90	1.4587	1.1297	2.617
	5	2.80	0.9162	0.9712	1.931
	20	2.40	0.9219	0.9688	1.918
2D-Q4+	5	2.80	0.8109	0.9556	1.931
	20	2.40	0.8655	0.9761	1.918
2D-Q8	5	2.80	0.8845	0.9977	1.931

[a]This is a limiting case. With only one element, the disk has a simply tapered cross section and not a hyperbola.

7.11 SUMMARY

The original title for this chapter was "Properties of the Elements." However, this sort of a title is misleading because the behavior of an element is not only a function of the element itself, but of equal importance are the type of loading, type of constraints, and type and degree of distortion. A review of the various cases shown here with the same elements shows that it is not possible to publish sets of accuracy numbers for specific elements, isolated from the model to which they will be applied.

The proposed set of benchmark cases by MacNeal and Harder has provided finite element users with an excellent set of simple standards against which to check various finite-element codes and elements. The value of the work published by MacNeal and Harder is not so much the actual performance numbers on the elements that they tested, but rather the setting forth of a simple, yet realistic set of problems against which any user can test his code.

When a finite element user has a new problem to model and is faced with the step of selecting an element type(s) and element density to use, it is recommended that one of the benchmark problems, which most closely represents the actual case, be selected and simple models with variable element densities and element types be run. The published performance numbers can be used for guidance, but the user should repeat the benchmark runs with his particular code, version, and selected elements to investigate the element performance. The problems in the benchmark set are simple enough that a case can be set up and run for a series of element densities with a minimal amount of labor and computer time.

Two other excellent sources of potential benchmark cases are Roark's *Formulas for Stress and Strain* [6] and Timoshenko's *Strength of Materials* [8]. These two sources offer the best collections of closed-form solutions to practical solid mechanics problems against which to test finite-element models.

REFERENCES

1. MacNeal, R. H. and R. L. Harder, "A Proposed Set of Problems to Test Finite Element Accuracy," *Finite Elements in Analysis and Design*, No. I, North-Holland, Amsterdam, 1985.
2. Timoshenko, S. and S. Woinowsky-Krieger, *Theory of Plates and Shells*, 2nd ed., McGraw-Hill, New York, 1959, pp. 120, 143, 202, 206.
3. Scordelis, A. C. and K. S. Lo, "Computer Analysis of Cylindrical Shells," *Journal of the American Concrete Institute*, No. 61, 1969, pp. 539–561.
4. Zienkiewicz, O. C., *The Finite Element Method*, McGraw-Hill, London, 1977, p. 417.
5. Morely, L. S. D. and A. J. Morris, "Conflict Between Finite Elements and Shell Theory," Royal Aircraft Establishment Report, London, 1978.
6. Roark, R. J. and W. C. Young, *Formulas for Stress and Strain*, 5th Ed., McGraw-Hill, New York, 1975.
7. Stodola, A., *Dampf- und Gasturbinen* (Steam and Gas Turbines), 6th ed., 1924, pp. 312–340.
8. Timoshenko, S., *Strength of Materials*, Part II, 3rd ed., Van Nostrand, New York.

8

ELEMENT DISTORTION

Finite element models of geometrically complex structures have the potential problem of element distortion. It is known that elements give the best results when they are used in their ideal shape, i.e., equilateral triangles, square quadrilaterals, and three-dimensional solid elements that are perfect cubes. In practice, however, it is nearly impossible to have all elements with perfect shapes. Elements will tolerate a modest amount of deviation from their ideal shape without any noticeable decrease in accuracy. Beyond a certain point of distortion, element stiffness and stress accuracy drop off dramatically. What constitutes an acceptable amount of distortion and what defines the point of accuracy rolloff are difficult questions whose answers will be different for every combination of element type, constraints, and loading. This determination is not apparent in a casual examination of any given element derivation and requires a detailed theoretical analysis beyond the patience of most finite element users. The best discussions of the effects of distortion on various element types are given by Cook

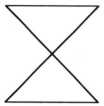

Figure 8.1 Element distortion: Crossed numbering.

(1, Chapter 5) and by Bathe and Wilson (2, Chapter 4). The topic of distortion has already been addressed in the previous chapter in discussions of several of the test cases such as the twisted beam, the Scordelis--Lo roof, and the spherical shell.

There are several cases of gross element distortion listed in Chapter 5 and repeated here that no element can tolerate and that should result in a fatal error in any finite-element program.

Element numbering must be in a clockwise or counterclockwise fashion and never in an X pattern as shown in Figure 8.1.

Elements must never double back on themselves (Figure 8.2).

Element sides must never be collapsed so that the corner is close to 180° (Figure 8.3).

In many programs, elements may not have two or more coincident nodes as shown in Figure 8.4. It is extremely important to consult the user's manual for any program to determine whether or not the use of coincident or double nodes is acceptable.

It is never acceptable to have an element with zero area or volume (Figure 8.5).

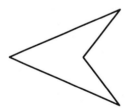

Figure 8.2 Element distortion: Doubled-back shape.

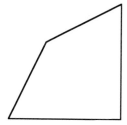

Figure 8.3 Element distortion: Collapsed element.

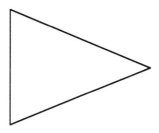

Figure 8.4 Element distortion: Coincident nodes.

Figure 8.5 Element distortion: Zero area.

All of the above conditions will result in an element with either zero or negative stiffness terms that will, in turn, cause a fatal error in either the program's element subroutine or global solution routine.

For the present purposes, element distortion may be broken down into five categories. MacNeal and Harder [3] discuss the element distortion of a simple beam and refer to four types of element distortion. Their name convention will be followed here

Aspect ratio The ratio an element length to width (or height for three-dimensional cases), the ideal ratio is 1.0 giving a square (Figure 8.6).

Taper The ratio of two opposite sides, an aspect ratio other than 1.0 gives a trapezoid shape (Figure 8.7).

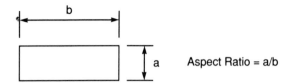

Aspect Ratio = a/b

Figure 8.6 Aspect ratio definition.

Skew angularity The element vertex angles, ideal is 90° for a quadrilateral and 60° for a triangle (Figure 8.8).

Warping Applies to plate elements, the measure of the distance that a node lies out of the plane defined by the other three nodes of a quadrilateral plate.

Bow Applies to plate elements that are curved and form part of a cylindrical surface. Bow refers to the ratio of the height of the arc to the chord length.

8.1 SIMPLE BEAM: ONE LAYER OF ELEMENTS

MacNeal and Harder [3] use a simple beam with three-element patterns to test the effects of moderate distortion on the accuracy of several element types. The beam is 0.2 wide, 0.1 thick, and 6.0 long. Young's modulus is $1.0E^7$ and the Poisson's ratio 0.3. The

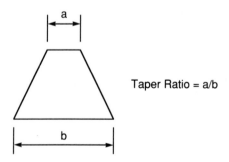

Taper Ratio = a/b

Figure 8.7 Taper ratio definition.

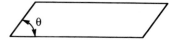

Figure 8.8 Skew angularity definition.

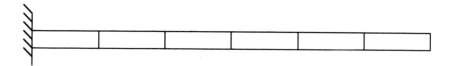

Figure 8.9 Rectangular beam: Baseline model.

three-element patterns are shown in Figures 8.9—8.11. The first used a single layer of rectangular elements across the section, the second adjusted the nodes to give a skew angle of 45° (parallelogram elements), and the third adjusted the nodes to give a taper ratio of 0.67 (trapezoidal elements).

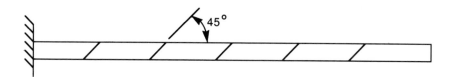

Figure 8.10 Rectangular beam: Parallelogram elements.

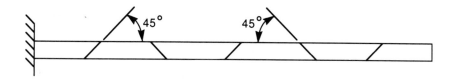

Figure 8.11 Rectangular beam: Trapezoidal elements.

Four loading conditions are applied, axial tension, shear loading at the tip in both the in-plane and out-of-plane directions to give a combined direct and bending loading, and a twist applied to the tip of the beam. Unit load amplitude is applied in all cases. Theoretical tip displacements are obtained from beam theory

Tension	tip displacement = 3×10^{-5}
In plane	tip displacement = 0.1081
Out of plane	tip displacement = 0.4321
Twist	tip displacement = 0.03208

The results of the MacNeal—Harder study along with results from two additional three-dimensional solid elements are given in Table 8.1. The results show good performance for all elements in the tensile loading case. For a simple geometry and uniaxial tensile loading, it is expected that all elements should give good results. For the cases of in-plane and out-of-plane shear loading at the tip, the results are generally good with a few exceptions. The LD plate (P-Q4) element did not do well with in-plane loading when the elements are distorted in the parallelogram and trapezoidal shapes. This is not a problem with the QD plate (P-Q8) as it gave consistently good performance for all three gometries and load conditions. The QD plate can support a linear stress distribution across its section.

8.2 EFFECT OF NUMERICAL INTEGRATION ORDER

The three LDADS three-dimensional solid elements used in the simple beam test case and shown in Table 8.1 gave different results. Although all three of these elements can be classified as LDADS three-dimensional solid elements, there are differences in the numerical integration procedures used to produce the stiffness matrices. In forming the element stiffness matrix, it is required to integrate element stiffness over the element volume. The common numerical integration techniques, such as Gauss quadrature or the Newton-Cotes formula, requires that a stiffness function be evaluated at a number of points within the element volume. The values of this function are then multiplied by appropriate weighting factors and summed to give the value of the integral.

The order of numerical integration, i.e., the number of points to be used, may be selected to optimize the accuracy vs. computer

Table 8.1 Results of Straight Beam Test Cases (six element types, 6 × 1 element grid pattern)

Rectangular elements	Normalized displacements					
	P-Q4	P-Q8	3D-H8+	3D-H8+(S)	3D-H8+(F)	3D-H20
Tension	0.995	0.999	0.978	0.988	0.978	0.994
In plane	0.904	0.987	0.978	0.981	0.978	0.970
Out of plane	0.986	0.991	0.981	0.981	0.973	0.961
Twist	0.941	0.950		0.910		0.904

Parallelogram elements	Normalized displacements					
	P-Q4	P-Q8	3D-H8+	3D-H8+(S)	3D-H8+(F)	3D-H20
Tension	0.996	0.999	0.970	0.989	1.007	0.994
In plane	0.080	0.995	0.666	0.080	0.638	0.967
Out of plane	0.977	0.985	0.562	0.055	0.544	0.941
Twist	0.945	0.965		0.904		0.913

Trapezoidal elements	Normalized displacements					
	P-Q4	P-Q8	3D-H8+	3D-H8+(S)	3D-H8+(F)	3D-H20
Tension	0.996	0.999	1.015	0.989	1.007	0.994
In plane	0.071	0.946	0.097	0.069	0.655	0.964
Out of plane	0.968	0.998	0.061	0.051	0.045	0.964
Twist	0.951	0.943		0.906		0.904

time for the elements. Bathe and Wilson [2] point out that for a
polynomial of order n (n = 1 for a LD element, n = 2 for a QD
element), the Newton—Cotes formula requires n + 1 points and
Gauss quadrature integration requires 2n - 1 to exactly integrate the function. For example, a LD or LDADS three-dimensional solid element integrated by Gaussian quadrature would require a minimum of 2 × 2 × 2 integration points, but 3 × 3 × 3
integration points would give an exact integration. Bathe and
Wilson [2] and Cook [1] state that exact integration gives an
upperbound on stiffness, i.e., the finite-element method in-
herently overpredicts stiffness and, in general, converges to
an overstiff solution when exact numerical integration is used.
The use of reduced integration order, e.g., 2 × 2 × 2 integration of a three-dimensional solid element rather than 3 × 3 × 3,
generally gives reduced stiffness. In many cases, the use of a
reduced integration order can be used to compensate for the inherent overstiffness of the element and give superior results.

Reduced integration order has the second benefit of substantially reducing computer time. Element formulation times and element stress recovery times are virtually a linear function of the
number of integration points for Guassian quadrature, so that
the difference between 2 × 2 × 2 (8 points) and 3 × 3 × 3 (27
points) is a factor of 8/27 or 0.296. When the time for equation
solution and overhead routines are included, the total difference
for the analysis is about 1:2.

Reduced integration order, however, has the potential disadvantage of producing erratic results in cases where the elements
are distorted to any degree. Reduced integration order may also
cause a singularity in the stiffness matrix that can, in turn, produce useless results. Elements formed with reduced integration
order, even when distorted, may produce acceptable results, according to Cook [1], when the element happens to be aligned with
a strain field that it can exactly represent. This phenomenon is
illustrated by the three three-dimensional solid elements in Table
8.2. The three LDADS three-dimensional solid elements are

> 3D-H8+(R) three-dimensional solid element, extra bending
> shapes, 2 × 2 × 2 reduced integration.
> 3D-H8+(S) three-dimensional solid element, extra bending
> shapes, selected reduced integration.
> 3D-H8+(F) three-dimensional solid element, extra bending
> shapes, full 3 × 3 × 3 integration.

In addition, two QD elements, 3D-H20 and 3D-H20(R), are included in Table 8.2. The 3D-H20 element is the 20-node, QD,

Table 8.2 Results of Straight Beam Test Cases (five element types, 6 × 1 element grid pattern)

Square elements[a]	Normalized displacements[a]	
	3D-H8+	3D-H20
Tension	0.988	1.250
In plane	0.994	0.988
Out of plane	0.994	0.988

Rectangular elements	Normalized displacements				
	3D-H8+	3D-H8+(S)	3D-H8+(F)	3D-H20	3D-H20(R)
Tension	0.978	0.988	0.978	0.994	0.999
In plane	0.978	0.981	0.978	0.970	0.984
Out of plane	0.981	0.981	0.973	0.961	0.972
Twist		0.910		0.904	0.911

Parallelogram elements	Normalized displacements				
	3D-H8+	3D-H8+(S)	3D-H8+(F)	3D-H20	3D-H20(R)
Tension	0.970	0.989	1.007	0.994	0.999
In plane	0.666	0.080	0.638	0.967	0.994
Out of plane	0.562	0.055	0.544	0.941	0.961
Twist		0.904		0.913	0.918

Trapezoidal elements	Normalized displacements				
	3D-H8+	3D-H8+(S)	3D-H8+(F)	3D-H20	3D-H20(R)
Tension	1.015	0.989	1.007	0.994	0.999
In plane	0.097	0.069	0.655	0.964	0.964
Out of plane	0.061	0.051	0.045	0.964	0.964
Twist		0.906		0.904	0.918

[a]From baseline case, Table 7.5.

three-dimensional solid, isoparametric solid with full 3 × 3 × 3 integration. The 3D-H20(R) element is the 20-node, QD, three-dimensional solid with reduced integration.

Table 8.2 is a subset of Table 8.1 with the addition of the baseline performance data on the 3D-H8+(R) element taken from Table

7.5. In this case, there is no substantial difference in accuracy between the perfectly cubic elements of the baseline case from Table 7.5 and the rectangular elements with a length-to-thickness aspect ratio of $0.1/1 = 0.1000$. This can be attributed to the fact that the results for both element shapes are close to the theoretical solution. All elements gave good results in simple tension regardless of the element shape. This is due to the fact that the elements are representing a constant, uniaxial stress field, and, as Cook points out, LD or LDADS elements should be able to represent a constant stress condition even with distortion. When the elements are called on to represent a stress gradient, especially a reversed stress gradient, the LDADS elements begin to show unacceptable results. The QD elements, by virtue of the fact that they can exactly represent a linear variation in strain and stress within the element, gave good results for these beam bending case with its fully reversed bending stresses.

For the parallelogram elements, the 3D-H8+(R) ($2 \times 2 \times 2$ integration) and 3D-H8+(F) ($3 \times 3 \times 3$ integration) did substantially better in bending than the LDADS three-dimensional solid element with selected reduced-order integration 3D-H8+(S) in the two bending cases. The two QD elements gave good results for all load cases.

For the trapezoidal elements, there is a difference in the behavior for in-plane bending between the 8-node solid element with full $3 \times 3 \times 3$ integration [3D-H8+(F)] and the two 8-node solid elements with reduced integration. For the loading case of out-of-plane bending, all three LDADS elements gave equally poor results.

The 20-node solid element with the reduced integration order gave slightly better results than the same element using full $3 \times 3 \times 3$ integration. All results for the QD elements, and for the most part for all of the elements, showed that the elements are overstiff, as evidenced by most of the ratios being less than 1.0 in Table 8.2. The fact that there is an increase in flexibility with the QD solid element using a reduced integration scheme verifies Cook's, Bathe and Wilson's statements that the increased flexibility of less than exact integration helps to balance the inherent overstiffness of the elements.

8.3 SIMPLE BEAM: THREE-ELEMENT LAYERS

A second beam model was set up consisting of three layers of three-dimensional solid elements as shown in Figure 8.12 in order

Completely Constrained

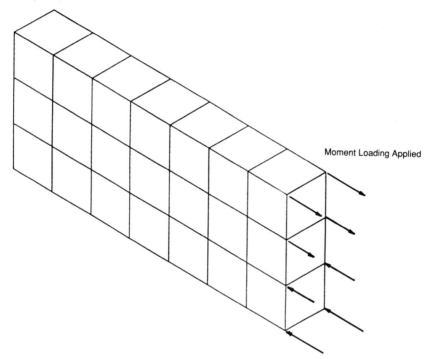

Moment Loading Applied

Figure 8.12 Three-layer beam: Baseline model.

to study the combined effects of element skew angularity and aspect ratio, and the effect of taper ratio. The beam dimensions are 1 × 3 × 7. The Young's modulus is 30,000,000 and Poisson's ratio 0.3. The loading consisted of a set of forces, as shown in Figure 8.12, giving a bending moment at the free end.

Four cases of skew angle, including a baseline case of 90°, are run as follows: 90, 45, 30, and 15°. Four cases of taper ratio, including a baseline case of taper ratio equal to 1.0 are run as follows: 1.00, 0.200, 0.02, and 0.002. For the taper ratio investigation, three different elements are used: 3D-H8+(F), 3D-H8+(R), 3D-H8. In the skew angularity test, two element types, 3D-H8+ and 3D-H8(R), are used.

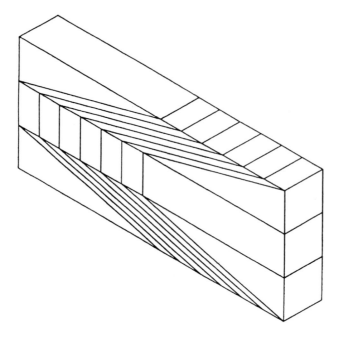

Figure 8.13 Three-layer beam: Skew angle/aspect ratio test.

8.3.1 Skew Angle/Aspect Ratio Test

The vertex angles for the top and bottom layer of elements are varied through 90 (perfect), 45, 30, and 15° as shown in Figure 8.13. The results are given in Tables 8.3 for displacements and 8.4 for stresses. These results showed that for the 3D-H8+(F) element, the accuracy did not drop off substantially until the skew angle is greater than 60° and the width-to-length ratio is less than 0.22. The LD element 3D-H8(F) showed a uniform dropoff in accuracy with a substantial drop from 0.683 to 0.270 in normalized stress, going from 60 to 75°.

In these cases of the three-layer beam, the accuracy of stresses at the surface was investigated as well as global displacements. In actual practice, not all elements in a model will have the same degree of distortion as they do in the single-layer beam model. Most of the elements in a typical model will not be badly distorted. However, groups of elements in transition areas may

Table 8.3 Results of Straight Beam Distortion Test, Displacement Results (three element types, 6 × 3 element grid pattern, in-plane moment loading)

Taper test			
	Normalized displacements		
Taper ratio	3D-H8(F)	3D-H8+(F)	3D-H8+(R)
1.000	0.950	1.000	1.000
0.200	0.926	0.999	1.000
0.020	0.790	0.990	1.000
0.002	0.631	0.985	1.000

Skew test		
	Normalized displacements	
Skew angle	3D-H8(F)	3D-H8+(F)
90.0	1.000	1.000
45.0	0.729	0.945
30.0	0.521	0.851
15.0	0.282	0.725

have to be distorted to accommodate element patterns or model geometry. In these cases, the global displacements may not be substantially affected by moderate inaccuracy in the stiffness of a few distorted elements. However, the stress solution may produce stress results in those distorted elements that have a moderate error. Although this may be a localized effect in the model, it may be disconcerting in any event.

8.3.2 Taper Ratio Test

The results of the taper ratio study for taper ratios of 1.0 (perfect), 0.20, 0.02, and 0.002 are shown in Tables 8.3 and 8.4 for displacements and stresses, respectively. The model for the taper ratio of 0.02 is shown in Figure 8.14. Displacement results showed that both formulations of the LDADS element gave good results with the maximum error of only 1.5% for the LDADS element with full 3 × 3 × 3 integration [3D-H8+(F)] and a taper ratio

Table 8.4 Results of Straight Beam Distortion Test, Stress Results (three element types, 6 × 3 element grid pattern, in-plane moment loading)

	Taper test		
	Normalized stresses		
Taper ratio	3D-H8(F)	3D-H8+(F)	3D-H8+(R)
1.000	0.995	1.000	1.000
0.200	1.133	1.090	1.016
0.020	1.189	1.149	1.035
0.002	1.219	1.152	1.035

	Skew test	
	Normalized stresses	
Skew angle	3D-H8(F)	3D-H8+(F)
90.0	0.995	1.000
45.0	0.840	1.003
30.0	0.683	1.012
15.0	0.270	1.140

of 0.002, i.e., effectively wedges on the top and bottom layer of elements. The element with full integration underpredicted displacement, whereas the element with reduced integration [3D-H8+(R)] gave perfect displacements, to four significant digits, for all taper ratio cases. This shows, again, that the increased flexibility of reduced integration countered the inherent overstiffness of the element. The linear element (3D-H8), as expected, is overstiff and the most sensitive to taper ratio.

The stress results showed trends that are similar to those of the displacement results. The LDADS element with reduced integration performed the best of the three elements, followed by the LDADS element with full integration and the LD element. All elements overpredicted stress. This may be explained as follows: (1) the effect of a nonzero taper ratio is to make the element overstiff, (2) this overstiffness is moderated by the center row of elements that have more accurate and flexible stiffness properties, (3) the global displacements are not as small as they would have been if all elements in the beam had the overstiff properties

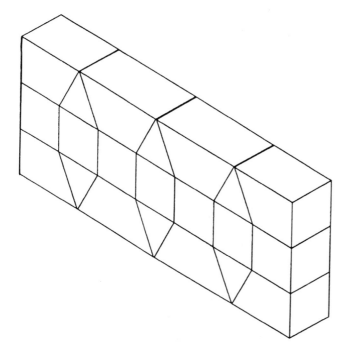

Figure 8.14 Three-layer beam: Taper ratio test.

of the top and bottom layer, (4) the global displacements combined with the higher stiffness and stress properties give higher stresses.

The LDADS elements showed a leveling off of error below a taper ratio of 0.02. Changing the taper ratio an order of magnitude from 0.02 to 0.002 produced only a 0.3% difference in stress results for the LDADS element with full integration [3D-H8+(F)]. There is no change in stress for the LDADS element with reduced integration [3D-H8+(R)] for taper ratios of 0.02 and 0.002. The maximum error for this element is only 3.5% at taper ratios of 0.02 and 0.002.

8.4 SUMMARY

The accuracy of general element types cannot be given in terms of simple element distortion parameters alone. Factors that influence element accuracy are

Type of element formulation Whether or not the element is iso-parametric, the order of the element, i.e., LD, LDADS, or QD, and the number of integration points, i.e., 8 (2 × 2 × 2) or 27 (3 × 3 × 3) Gauss points.

Element distortion Aspect ratio, taper ratio, skew angularity, and twist of a solid element.

Type of loading Tension, bending, torsion.

Accuracy depends on the interaction of all of the above parameters and it is not possible to "decouple" them to quantify their individual effects.

Some general guidelines can be offered to help one avoid serious element distortion problems

The *aspect ratio* should be no more than 10:1 in noncritical areas and 2:1 in elements where stresses are to be obtained such as notches and fillets.

The *taper ratio* should be at least 0.10 in noncritical areas and 0.25 in critical elements.

The element *vertex angles* should not be less than 45° for quadrilateral elements and 15° for triangular elements.

LD and LDADS elements do not like stress gradients. Even the LDADS elements are basically constant strain elements. Particular care should be paid to single layers of LD or LDADS elements. If possible, a double layer of elements should be used if the area is in bending.

To interpret element stress data and check for high-stress gradients, use the stresses at the Gauss points. This will allow for 8 or 27 stress points to be examined for each selected three-dimensional solid element. A large gradient across the Gauss point stresses indicates the potential for bad stress results, especially if there is a reversal of sign within the elements. If such a gradient was to exist, it could be corrected by increasing the element density or using a higher-order element.

REFERENCES

1. Cook, R. D., *Concepts and Applications of Finite Element Analysis*, Wiley, New York, 1974.
2. Bathe, K. J. and E. L. Wilson, *Numerical Methods in Finite Element Analysis*, Prentice-Hall, Englewood Cliffs, N.J., 1976.

3. MacNeal, R. H. and R. L. Harder, "A Proposed Set of Problems to Test Finite Element Accuracy," *Finite Elements in Analysis and Design*, No. I, North-Holland, Amsterdam, 1985.

9

REFINED MESH MODELING

In developing a finite element analysis model, the engineer is always faced with the problem of trading off accuracy against computer runtime. Because the finite element method is a numerical approximation, there is no absolutely correct answer; the accuracy generally increases with the number of elements used. Although there is a point of convergence, beyond which an increase in elements will not substantially increase accuracy, this point usually requires thousands of elements for a large, complicated structure. There is a real problem when stresses are highly concentrated at locations such as bolt holes, fillets, or welds, etc. In these cases, the entire structure may have to be included in the model in order to properly determine the maximum stresses.

Stress result accuracy varies as a function of element density for two principal reasons: (1) A finer mesh allows for better definition of the component geometry details, and (2) a finer mesh allows more stress calculation points to define the stress gradient in a notch or fillet region. Stresses are calculated either at the element centroid, Gaussian integration points, or nodes (and sometimes at midside surface locations). In general, element centroidal stresses are preferred because they are based on nodal displacements that bound the centroid. Nodal stresses at a model boundary must be calculated based on displacements that do not completely bound the nodal point and often represent stresses from more than one element, these stresses being extrapolated rather than interpolated. In calculating the stress in a fillet, for example, the engineer intuitively knows that the maximum stress will occur at the surface. The engineer must make the decision as to whether to rely on the element centroidal stresses that are not calculated at the surface or to use the nodal stresses that may be extrapolated based on only one element and may not represent the true stress gradient near the surface.

9.1 REFINED MODELING CONCEPT

The alternatives for modeling a complex structure such as this are the following:

1. *Coarse model* A model that is adequate to characterize the response of the structure, either statically or dynamically, and the stress concentration effects are ignored. Maximum stresses must be calculated by ratioing the finite element analysis stress results with a suitable stress concentration factor taken from a source such as Peterson [1].

2. *Fine model* A model is generated with a gridwork suitable for determining the concentrated stresses throughout the model. This means overkill in the noncritical regions. This may be no more time-consuming to generate than the coarse model, but the computer runtimes are excessively high.

3. *Transition model* The critical areas of the structure are modeled with a suitable number of elements to achieve a good approximation of the peak stresses and stress gradients. The element density is reduced away from the critical areas by using a series of transition layers, i.e., 3-node triangular elements for two-dimensional models and 6-node wedge elements for three-dimensional models. This method has the disadvantage of requiring substantially higher modeling effort in order to achieve a good mesh without element distortion. It has the other disadvantage of element incompatibility in certain finite element programs. For example, some 3-node elements do not behave the same as 4-node elements.

4. *Refined mesh modeling* Also known as submodeling, this method requires two, separate models to be developed. The global model is a coarse model, as described above, and the second model is a fine mesh of only the critical region. Boundaries of the refined model will consist of free surfaces and cut boundary surfaces across which stresses or displacements from the global model need to be applied. Because both models can be relatively straightforward, the total effort involved to generate both may be less than the effort required to generate a transition model.

The refined model concept has several advantages. A relatively coarse model can be used to characterize the overall structural response. Extremely fine gridworks can be used to calculate the maximum stresses. A two-dimensional refined model may be used with a three-dimensional plate or solid global model, provided, of course, that there are no out-of-plane displacements in the critical regions. A two-dimensional model allows for about three times as many two-dimensional than three-dimensional elements for the same computer runtime and also allows for gap effects to be efficiently incorporated. In the event that an anticipated area of high stress is found with the global model, a second refined model can be generated for the second area. This could not be done with a transition model.

Transferring boundary conditions from the global to refined model is the most complicated part of using a refined model. Boundary conditions can be transferred as either interpolated displacements from the global model or as applied loading, i.e., stresses from the global analysis translated into pressure loadings for the refined model. The applied forces or displacements must be interpolated because there will always be more boundary nodes in the refined model than the global one. In forming the refined model, the boundary of the refined model should line up with selected element boundaries of the global model to make it easy to transfer displacements from the global element boundaries to the refined model boundary.

The use of a refined model is not a universal answer to every finite-element analysis. There is a requirement that the global model be detailed enough to give an accurate displacement or stress field away from the stress concentration site. Conversely, the stress concentration site must be small enough, relative to the total structure, so as not to influence the overall structural response. The cut boundaries of the refined model must be positioned far enough away from the stress concentration site so that the stress or displacement field of the coarse model, without this stress concentration effect, will be an accurate representation of the actual case at those locations, i.e., St. Venet's principal.

Some guidelines for refined mesh modeling are offered in Table 9.1.

Table 9.1 Rules of Thumb for Refined Mesh Modeling

— The global model must be of sufficient accuracy to provide good results at the cut boundary locations. In other words, the amount of inaccuracy at the cut boundary locations cannot be improved upon in the refined model.

— The cut boundary locations in the global model should be in regions of relatively flat, or at least linear, displacement gradients. If these locations cannot be determined beforehand, then the refined model should not be generated until the results of the global model are available.

— Boundary conditions may be transferred from the global model to the refined model either as displacements or as stresses from the global model applied as a distributed loading to the refined model boundary. Care must be taken when interpolating displacements for a refined model boundary. Small differences in displacements can result in substantial strains. In many programs, it is easier to apply multiple load steps using applied forces rather than applied displacements.

— The shape functions of the global model, or their equivalent, should be used to interpolate displacements or stresses for application to the refined model boundaries.

— The stress gradients from the refined model should be examined and compared with the stress gradients of the global model. If there is any question as to the true maximum stress in the refined model, the refined model may be regenerated using an even finer mesh or a second refined model, i.e., a portion of the original refined model, can be generated.

Constrained

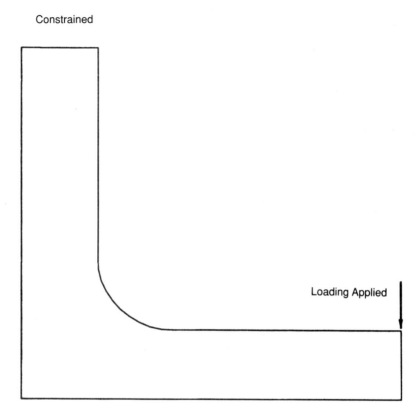

Figure 9.1 Right-angle beam with corner fillet.

9.2 RIGHT-ANGLE BEAM EXAMPLE CASE

The following example shows a comparison of four modeling techniques used for a simple beam structure with a right angle and radiused corner fillet (Figure 9.1). This example illustrates the procedure of refined mesh modeling, its strengths, and some significant drawbacks.

The four models are (1) a fine model with 168 elements, (2) a transition model with 148 elements, (3) three coarse models with 20 and 21 elements, and (4) a refined model of the fillet area with 72 elements.

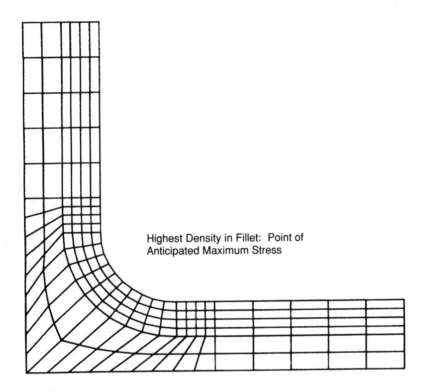

Highest Density in Fillet: Point of
Anticipated Maximum Stress

Figure 9.2 Right-angle beam, fine model, 168 elements.

The fine model (Figure 9.2) represents the corner fillet by using 10 rows of elements to represent the fillet radius and four layers of elements evenly spaced to pick up the stress gradient into the depth of the beam for a total of 168 elements, 203 nodes, and 392 active degrees of freedom. The computer runtime for this and the other models is given in Table 9.2. As expected, the point of maximum calculated stress was in the fillet radius, as shown in Figure 9.3. The maximum calculated stresses are given in Table 9.3.

A transition model was developed having the same grid density in the fillet region, but a reduced density gridwork at the noncritical ends of the beam (Figure 9.4). The transition from

Table 9.2 Details and Computer Runtimes for Three Right-Angle Beam Models

---

Coarse models
    20 elements (two-dimensional, plane stress)
    33 Nodes   60 active DOF
    Times (sec)

Element formulation	3.218 (0.161 sec/elem.)
Wavefront solution	1.515
Stress solution	1.715 (0.086 sec/elem.)
Element forces	0.127
Misc.	6.216
Total	12.791 sec

Transition model
    148 Elements (two-dimensional, plane stress)
    165 Nodes   328 active DOF
    Times (sec)

Element formulation	23.088 (0.156 sec/elem.)
Wavefront solution	8.133
Stress solution	12.318 (0.083 sec/elem.)
Element forces	0.918
Misc.	20.476
Total	64.933 sec

Fine model
    168 Elements (two-dimensional, plane stress)
    203 Nodes   392 active DOF
    Times (sec)

Element formulation	27.412 (0.163 sec/elem.)
Wavefront solution	9.473
Stress solution	14.021 (0.083 sec/elem.)
Element forces	1.042
Misc.	12.416
Total	73.742 sec

Note: Computer runtimes are valid for a relative comparison between models and cases.

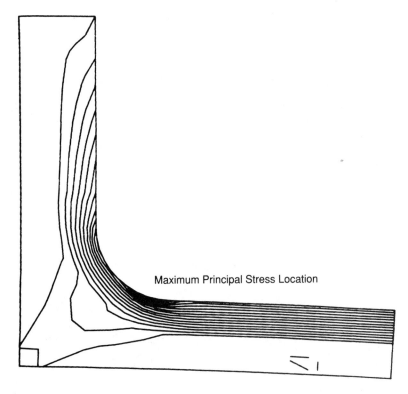

Figure 9.3 Stress contour plot, fine model, maximum principal stress.

the detailed to coarse gridwork was accomplished by a pattern of triangular elements that reduce by half the number of elements per row. The comparison of computer runtimes shows that the transition model required 88% of the computer time of the fine model and gave comparable results, especially the maximum nodal stresses. The difference in stresses is probably due more to using triangular elements in the model than to the difference in the number of elements. It must be noted that the transition model because of its nonuniformity requires more effort to create the model and input the data than the fine model, and it does not lend itself as easily to automatic mesh generation.

Table 9.3 Stress Results of Right-Angle Beam: Coarse, Transition, and Fine Models

	Maximum centroidal stress				Maximum nodal stress			
	σ_x	σ_y	τ_{xy}	σ_1	σ_x	σ_y	τ_{xy}	σ_1
Fine model	750.43	78.409	-145.34	780.51	946.39	59.46	-138.54	971.00
Transition model	792.92	45.601	-69.94	804.36	950.41	59.95	-137.13	974.69
Coarse model 1, no fillet radius	437.66	99.09	-27.39	439.86	851.54	-51.09	-7.572	852.02
Coarse model 2, additional element in corner	438.56	49.05	-3.64	438.59	850.21	-27.33	-0.887	850.87
Coarse model 3, with fillet radius	540.99	60.79	2.51	541.01	842.41	-41.45	-0.18	842.42

σ_x = X direction (horizontal axis) stress
σ_y = Y direction (vertical axis) stress
τ_{xy} = shear stress
σ_1 = maximum principal stress

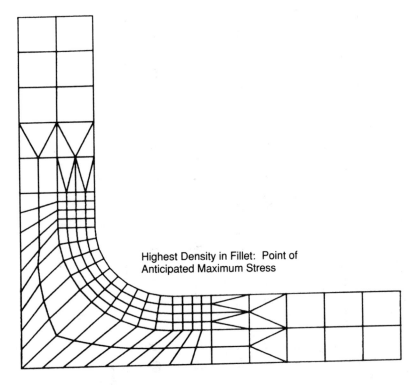

Figure 9.4 Right-angle beam, transition model, 148 elements.

Three coarse models were used in this example (Figures 9.5–9.7). It can be seen that the first coarse model does not even attempt to represent the corner fillet and simply models a square corner. A comparison of this model with the actual structure, shown in Figure 9.1, indicates that there is a substantial amount of material in the fillet that the first coarse model does not include. The first coarse model uses 20 elements, 33 nodes, and 60 active degrees of freedom. The second coarse model used one triangular element to attempt to represent the stiffness of the corner of the beam. The third coarse model is shown in Figure 9.7 and gives a representation of the fillet. This model also uses 20 elements with 33 nodes and 60 active degrees of freedom.

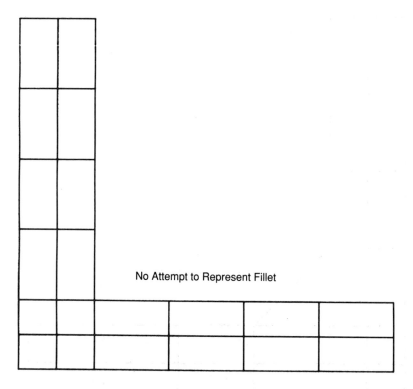

Figure 9.5 Right-angle beam, first coarse model, 20 elements.

The refined model is shown in Figure 9.8. It uses 72 elements and 95 nodes and has 136 active degrees of freedom. The refined model is a subset of the fine model and uses the fine model's gridwork in the fillet region. The refined model replaces approximately three elements of the first coarse model, and four elements of the second and third coarse model with 72 elements. Computer runtimes are given in Table 9.4 for the models.

Static displacements and stresses were calculated with the three coarse models and are shown in Figures 9.9–9.11 and Table 9.5. There was a significant difference in peak stress between the coarse models, as may be expected. The first

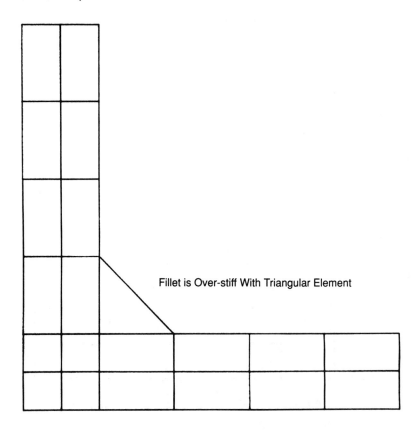

Fillet is Over-stiff With Triangular Element

Figure 9.6 Right-angle beam, second coarse model, 21 elements.

coarse model, which does not model the fillet, gave a maximum principal nodal stress of 852.02. Its maximum centroidal stress was only 439.86. Due to such a few elements, the element centroid was only half-way between the beam surface and neutral axis. The location of the maximum stress in the first coarse model was away from the corner, on the top surface of the horizontal leg of the beam. The second coarse model, which differs from the first only by the addition of a triangular element to represent the fillet, had a maximum principal nodal stress of 851.44 in the same location. The third coarse model that includes a fillet radius modeled with four elements had a maximum

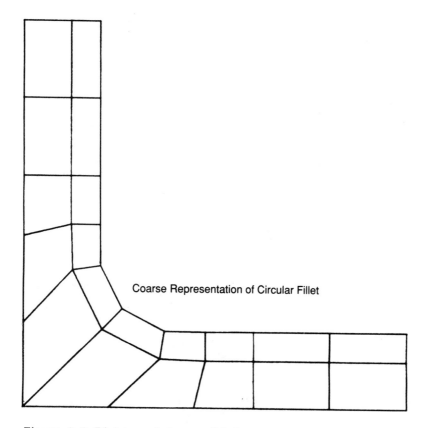

Coarse Representation of Circular Fillet

Figure 9.7 Right-angle beam, third coarse model, 20 elements.

nodal principal stress of 842.42 in the fillet. Comparing the three
maximum stress numbers from the coarse models against the maximum principal nodel stress from the fine model (971.00) shows
that even the third coarse model does not give results that are
better than 87% of the fine model's result. The third coarse
model did, however, give the correct location of maximum principal stress.

The results of the corresponding refined models are given in
Table 9.5 and Figures 9.12–9.14 based on displacements from
the three coarse models. The same refined model was used for
all three cases, the only difference being the applied cut

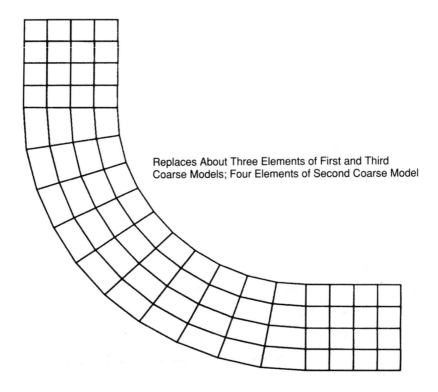

Replaces About Three Elements of First and Third
Coarse Models; Four Elements of Second Coarse Model

Figure 9.8 Right-angle beam, refined model of fillet area, 72 elements.

boundary displacements. The results showed that the displacements of the first coarse model, when used with the refined model, overpredicted the maximum principal stress based on the results of the fine model. This is due to the fact that the first coarse model overpredicts the displacements at the cut boundary locations because of the lack of material in the fillet and the resulting underprediction of stiffness in the fillet. The second coarse model's displacements underpredict stress because they underpredict the displacements at the cut boundary locations of the coarse model. Although there was a substantial difference between the nodal stresses of the refined model with the first

Table 9.4 Details and Computer Runtimes for Right-Angle Beam Refined Model

---

Coarse models
  20 Elements (two-dimensional, plane stress)
  33 Nodes   60 active DOF
  Times (sec)

Element formulation	3.218 (0.161 sec/elem.)
Wavefront solution	1.515
Stress solution	1.715 (0.086 sec/elem.)
Element forces	0.127
Misc.	6.216
Total	12.791 sec

Fine model
  168 Elements (two-dimensional, plane stress)
  203 Nodes   392 active DOF
  Times (sec)

Element formulation	27.412 (0.163 sec/elem.)
Wavefront solution	9.473
Stress solution	14.021 (0.083 sec/elem.)
Element forces	1.042
Misc.	12.416
Total	73.742 sec

Refined model
  72 Elements (two-dimensional, plane stress)
  95 Nodes   136 active DOF
  Times (sec)

Element formulation	11.888 (0.165 sec/elem.)
Wavefront solution	3.813
Stress solution	6.124 (0.085 sec/elem.)
Element forces	0.430
Misc.	13.430
Total	35.685 sec

Note: Computer runtimes are valid for a relative comparison between models and cases.

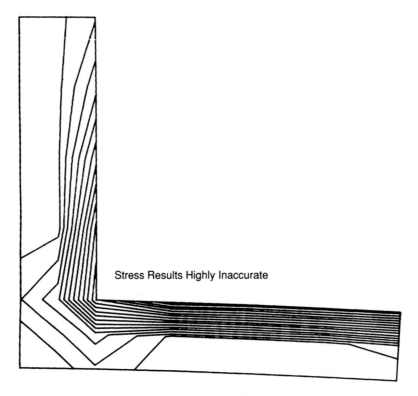

Figure 9.9 First coarse model, maximum principal stress.

and second coarse model displacement, the relative stress distributions are virtually identical as shown in Figures 9.12 and
9.13. Surprisingly, the results of the combination of the third
coarse model with the refined model did not significantly improve
the results, relative to the fine model.

Several variations were tried with the refined model to test
its sensitivity to modeling variables. The first test involved increasing the number of elements in the refined model by adding
16 additional elements (4 rows of 4) to the "top" of the model.
The purpose of this exercise is to test whether or not the cut
boundary is far enough from the fillet region so as not to be
influenced by the fillet. The results shown in Table 9.6 show

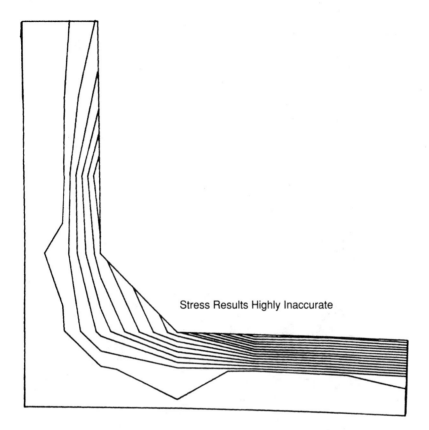

Figure 9.10 Second coarse model, maximum principal stress.

that there is some difference in maximum centroidal stress, but virtually no difference in maximum nodal stress.

The second test was aimed at determining the effect of accuracy of the interpolated, cut boundary displacements. Displacements along the top cut boundary were, intentionally, adjusted by 1% and the results are shown in Table 9.6. There was only about 0.5% difference in stresses.

The significant conclusions that can be drawn from this example case are that the coarse model has a significant influence of the end results and that a too coarse model will not give

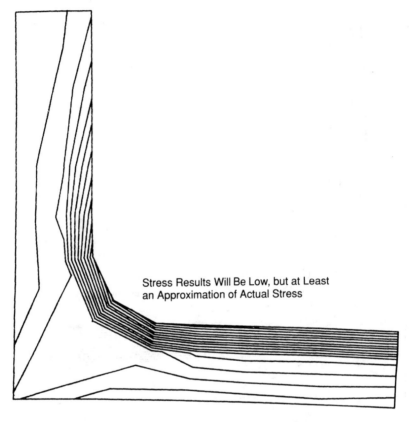

Stress Results Will Be Low, but at Least
an Approximation of Actual Stress

Figure 9.11 Third coarse model, maximum principal stress.

accurate results, regardless of how detailed the refined model is.
The use of the first coarse model with no representation of the
fillet radius was selected as a limiting case to determine the effect of using an extremely coarse model. The lack of material
in the fillet caused the first coarse model to have displacements
that were too high at the locations of the cut boundaries of the
refined model. When these displacements were transferred from
the coarse model to the boundaries of the refined model, the result was high stress in the refined model. In the case of this

Table 9.5 Stress Results of Right-Angle Beam: Coarse, Fine, and Refined Models

	Maximum centroidal stress				Maximum nodal stress			
	σ_x	σ_y	τ_{xy}	σ_1	σ_x	σ_y	τ_{xy}	σ_1
Fine model	750.43	78.409	-145.34	780.51	946.39	59.46	-138.54	971.00
Coarse model 1, no fillet radius	437.66	99.09	-27.39	439.86	851.54	-51.09	-7.572	852.02
Refined model with coarse model 1 displacements	856.25	67.47	-681.55	1454.7	893.38	858.01	-814.31	1696.0
Coarse model 2, additional element in corner	438.56	49.05	-3.64	438.59	850.21	-27.33	-0.887	850.87
Refined model with coarse model 2 displacements	593.59	-3.82	-23.87	594.55	727.90	-4.79	-30.16	724.19
Coarse model 3, with fillet radius	540.99	60.79	2.51	541.01	842.41	-41.45	-0.18	842.42
Refined model with coarse model 3 displacements	544.09	-4.05	-34.14	546.21	722.89	-4.78	-30.16	724.19

σ_x = X direction (horizontal axis) stress
σ_y = Y direction (vertical axis) stress
τ_{xy} = shear stress
σ_1 = maximum principal stress

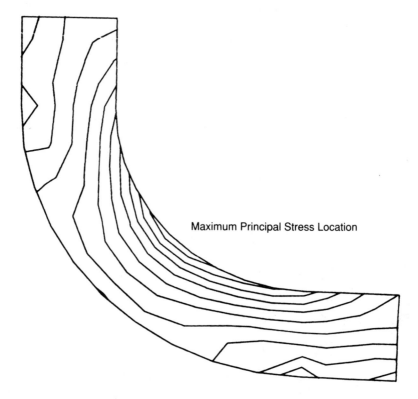

Maximum Principal Stress Location

Figure 9.12 Refined model, maximum principal stress: First coarse model displacements.

right angle beam, the details of the fillet geometry are not insignificant and need to be accurately included in the coarse model.

9.3 REFINED MODELING: EXAMPLE CASES

Several of the simple example cases of Chapter 5 will be used again to demonstrate the application of refined mesh modeling. Not all of the examples were chosen to show the positive side of this technique. Some of the example cases will show that

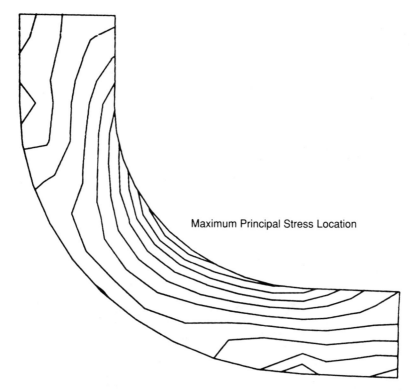

Maximum Principal Stress Location

Figure 9.13 Refined model, maximum principal stress: Second coarse model displacements.

refined mesh modeling should be selectively applied and is not the universal solution to obtaining optimum accuracy.

9.3.1 Example: Notched Block

The notched block two-dimensional model from Chapter 5, Section 5.4.2 is used here to illustrate the principals and procedures of refined mesh modeling. The two-dimensional global model is shown in Figure 9.15. This model has 38 two-dimensional plane strain elements, 49 nodes, and 82 active degrees of freedom. The global model uses 12 elements to represent the fillet area. This example, like the L-shaped beam, shows the gains in accuracy

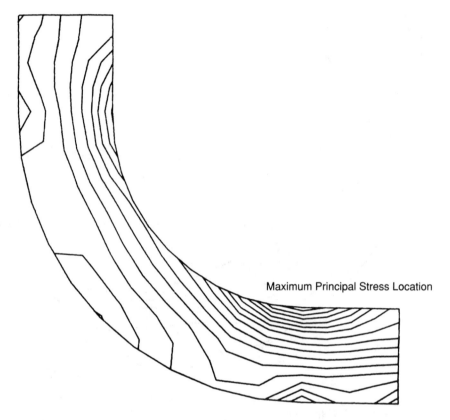

Figure 9.14 Refined model, maximum principal stress: Third
coarse model displacements.

of refined mesh modeling, but also points out the price in terms
of computer and modeling time.

The refined model is shown in Figure 9.16 and has a total of
64 elements, 166 nodes, and 280 active degrees of freedom. The
area of the refined model is approximately eight elements of the
global model. Computer runtimes for both models are given in
Table 9.7. As expected, the computer time for the refined model
was about double that of the global model because of the number
of elements and degrees of freedom in each.

Table 9.6 Right-Angle Beam: Variations of Refined Models

	Maximum centroidal stress				Maximum nodal stress			
	σ_x	σ_y	τ_{xy}	σ_1	σ_x	σ_y	τ_{xy}	σ_1
Fine model—reference	750.43	78.409	-145.34	780.51	946.39	59.46	-138.54	971.00
Refined model with coarse model 1 displacements	856.25	67.47	-681.55	1454.7	893.38	858.01	-814.31	1696.0
Refined model with coarse model 1 displacements modified 1%	853.40	674.33	-678.50	1448.2	890.16	851.30	-810.28	1686.9
Modified refined model with additional 16 elements with coarse model 1 displacements	689.05	798.85	-666.07	1412.0	892.40	855.91	-813.11	1693.2

σ_x = X direction (horizontal axis) stress
σ_y = Y direction (vertical axis) stress
τ_{xy} = Shear stress
σ_1 = Maximum principal stress

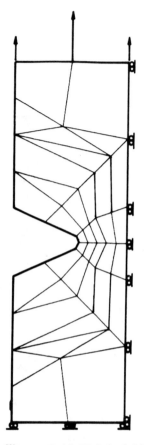

Figure 9.15 Notched block, two-dimensional global model.

The stress results presented in Table 9.8 showed that there was a moderate difference in the results between the global model and the refined model. The refined model gave higher stresses for the Y direction (the direction of load) and the maximum principal stress. A plot of the refined stress distribution is shown in Figure 9.17.

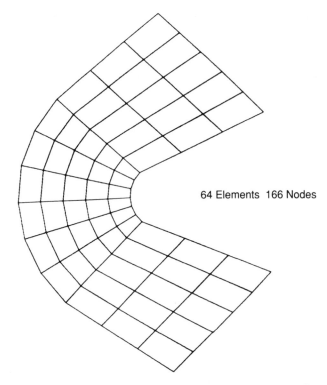

64 Elements 166 Nodes

Figure 9.16 Notched block, two-dimensional refined model.

In this case, the global model was not only sufficiently accurate, but it would appear that the use of a refined model with twice the number of elements and twice the computer runtime is hard to justify. A more cost-effective approach to this model might be to simply increase the number of elements in the fillet area to achieve a slightly better resolution of the stress field.

9.3.2 Example: Crank

The crank, discussed in Chapter 5, Section 5.4.4 and shown in Figure 9.18, represents a more practical application of refined

Table 9.7 Details and Computer Runtimes for Block Refined
Model

Two-dimensional global model
  38 Elements (two-dimensional, plane strain)
  49 Nodes   82 active DOF
  Times (sec)

Element formulation	8.149 (0.214 sec/elem.)
Wavefront solution	2.158
Stress solution	2.515 (0.066 sec/elem.)
Element forces	0.276
Misc.	3.510
Total	16.600 sec

Two-dimensional block refined model
  64 Elements (two-dimensional, plane strain)
  165 Nodes   280 active DOF
  Times (sec)

Element formulation	11.788 (0.184 sec/elem.)
Wavefront solution	3.491
Stress solution	5.412 (0.303 sec/elem.)
Element forces	0.406
Misc.	12.63
Total	33.73 sec

Note: Computer runtimes are valid for a relative comparison between models and cases.

mesh modeling in the keyway area. In the global model, three
elements are used to bound the keyway and there is no attempt
to model the stress field in the corner of the keyway. In the
global model, the key is not explicitly included in the model; the
nodes on the keyway that would contact the key are constrained
in the tangential direction to simulate the effect of the key bearing on the keyway.

A three-dimensional refined mesh model of the keyway region
is developed and shown in Figure 9.19. The refined model contains 168 three-dimensional solid elements (the same element used

Table 9.8 Stress Results of Notched Block Refined Model

Maximum centroidal stress				Maximum nodal stress			
σ_x	σ_y	σ_z	σ_1	σ_x	σ_y	σ_z	σ_1
Two-dimensional global model							
3.24	8.70	3.58	8.86	3.40	12.68	4.82	13.50
Two-dimensional refined model							
3.06	6.45	2.86	8.38	1.24	14.11	4.61	14.27

σ_x = X direction stress
σ_y = Y direction stress
τ_{xz} = shear stress
σ_1 = maximum principal stress

in the global model), 241 nodes, and 720 active degrees of freedom. The refined model uses 14 elements per layer to bound the keyway and positions one element at each corner to represent a small chamfer, which would normally occur in practice as the keyway is either broached or shaped.

The refined model had a free surface that was the inside of the keyway and three cut surfaces on the outside boundaries. There was one additional free surface of the refined model corresponding to the top of the crank and one additional cut surface at the "bottom" of the refined model. Three displacements were interpolated from the global model for each of the refined cut boundary nodes.

Computer runtimes are given in Table 9.8. Although the ratio of the number of elements between the two models was 1.33:1 and the ratio of the number of active degrees of freedom was 1.81:1, the ratio of total runtimes was 2.32:1. This is due to the decreased time for the frontal equation solution of the refined model that, in turn, is due to the more efficient element numbering, i.e., smaller wavefront, of the refined model.

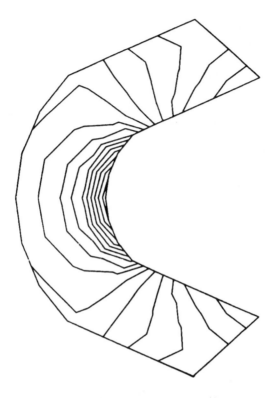

Figure 9.17 Notched block, two-dimensional refined model, maximum principal stress.

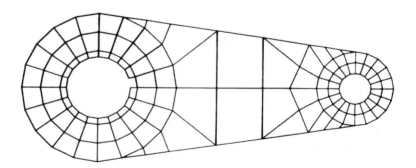

Figure 9.18 Crank, three-dimensional global model.

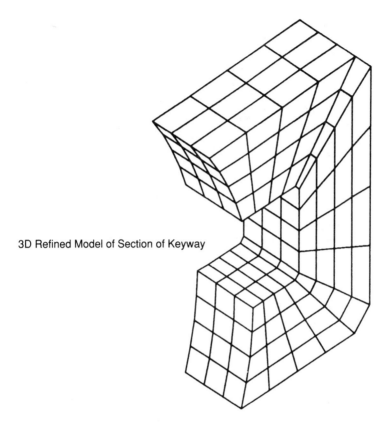

3D Refined Model of Section of Keyway

Figure 9.19 Crank, three-dimensional refined model of keyway.

The maximum principal stresses and the component stresses at the keyway are presented in Table 9.10. The comparison of maximum principal stresses shows that there is a 43% difference between the maximum centroidal stresses of the two models that is to be expected because the global model defines the keyway with a minimal three elements per layer and the refined model uses 14 elements per layer. Furthermore, the centroid of the refined model is much closer to the surface of the keyway than the centroid of the global model.

There was a smaller, 28% difference between the maximum principal nodel stresses of the two models, indicating that

Table 9.9 Details and Computer Runtimes for Crank Refined Model

Three-dimensional global model
  224 Elements (three-dimensional, solid)
  459 Nodes   1306 active DOF
  Times (sec)

Element formulation	320.53	(1.431 sec/elem.)
Wavefront solution	427.95	
Stress solution	77.609	(0.346 sec/elem.)
Element forces	5.821	
Misc.	37.418	
Total	924.38	sec

Three-dimensional refined model
  168 Elements (three-dimensional, solid)
  241 Nodes   720 active DOF
  Times (sec)

Element formulation	142.55	(0.849 sec/elem.)
Wavefront solution	156.25	
Stress solution	43.379	(0.258 sec/elem.)
Element forces	3.161	
Misc.	44.580	
Total	398.92	sec

Note: Computer runtimes are valid for a relative comparison between models and cases.

although the global model elements did a fair job extrapolating stresses to the node points, there is a limit to how much extrapolation these LDADS elements are capable of. The differences between the centroidal and nodal stresses of the two elements indicate that the LDADS elements are capable of estimating a stress gradient. The refined model, with its smaller elements, shows a smaller (0.9% compared to 13.2%) difference between the maximum centroidal stress and the maximum nodal stress.

A stress contour plot of the maximum principal stress in the global model around the keyway is shown in Figure 9.20 and a maximum principal stress plot for the refined model is shown in Figure 9.21.

Table 9.10 Stress Results of Crank Model

Maximum centroidal stress				Maximum nodal stress			
σ_r	σ_t	σ_a	σ_1	σ_r	σ_t	σ_a	σ_1
Global model, keyway							
-371.38	785.90	66.87	1126.8	-497.71	943.36	142.53	1274.43
Refined model, keyway							
68.27	1606.3	284.11	1615.3	-109.63	491.14	4.53	1630.22

σ_r = radial (relative to shaft bore) stress at keyway
σ_t = tangential stress at keyway
σ_a = axial stress at keyway
σ_1 = maximum principal stress at keyway

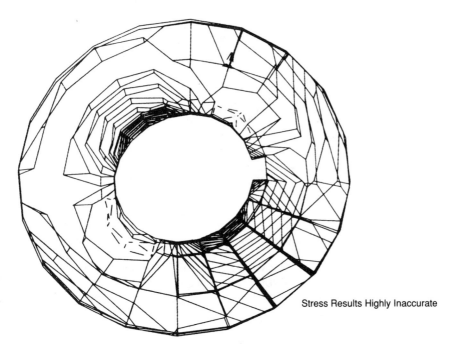

Stress Results Highly Inaccurate

Figure 9.20 Crank, global nodal keyway area, maximum principal stress.

Reasonable Maximum Stress Results

Figure 9.21 Crank, refined model keyway area, maximum principal stress.

In addition to calculating the maximum stresses, the refined model is necessary, in this case, to achieve any sort of reasonable definition of the stress field away from the surface of the keyway. This stress field definition would be necessary in cases of localized plasticity in the corners of the keyway or for fracture mechanics' applications for predicting crack growth.

9.3.3 Example: Gear Segment

A refined model of the fillet of the two-dimensional single gear tooth model, described in Chapter 5, Section 5.4.5. The global,

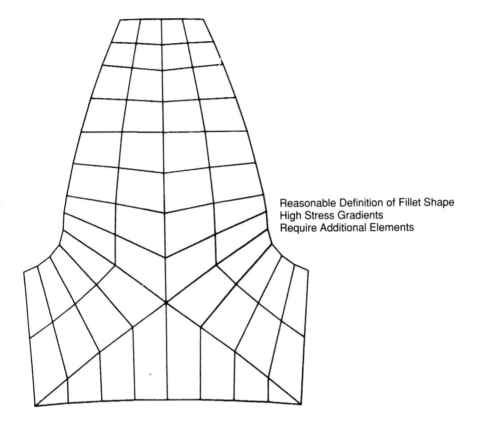

Reasonable Definition of Fillet Shape
High Stress Gradients
Require Additional Elements

Figure 9.22 Single gear tooth, two-dimensional global model.

two-dimensional single tooth model is shown in Figure 9.22 and
the refined model is shown in Figure 9.23. The area of the refined model covers five elements of the global model. The refined model has 60 elements, 80 nodes, and 112 active degrees of
freedom, virtually the same numbers as the global model. The
free surface of the refined model includes the fillet radius and
transition to the involute curve. The refined model has three
cut surfaces for which the nodal displacements were interpolated
from the global model results.

The computer runtimes are shown in Table 9.11. As expected,
the runtimes for the global model and refined model were very

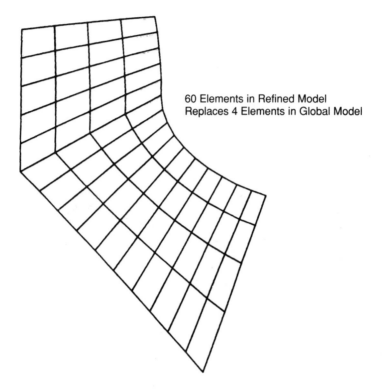

60 Elements in Refined Model
Replaces 4 Elements in Global Model

Figure 9.23 Single gear tooth, two-dimensional refined model of fillet.

close. Although both models have 60 nodes, the refined model has 112 active degrees of freedom as compared to 135 active degrees of freedom for the global model due to the number of imposed boundary displacements in the refined model.

Stress results based on the location of maximum principal stress are given in Table 9.12. These stresses show that the refined model stresses are consistently higher than the global model stresses for both centroidal and nodal stresses. Stress contour plots are shown in Figures 9.24 and 9.25 for the global and refined models, respectively. The plots show that the maximum stress occurs at the same position in both models. The

Table 9.11 Details and Computer Runtimes for Gear Refined
Model

Two-dimensional single tooth global model
  60 Elements (two-dimensional, plane strain)
  79 Nodes   135 active DOF
  Times (sec)
    Element formulation   13.14 (0.219 sec/elem.)
    Wavefront solution   3.767
    Stress solution   3.955 (0.066 sec/elem.)
    Element forces   0.445
    Misc.   13.03

    Total   34.337 sec

Two-dimensional fillet refined model
  60 Elements (two-dimensional, plane strain)
  80 Nodes   112 active DOF
  Times (sec)
    Element formulation   10.99 (0.183 sec/elem.)
    Wavefront solution   3.646
    Stress solution   5.079 (0.101 sec/elem.)
    Element forces   0.361
    Misc.   11.83

    Total   31.91 sec

Note: Computer runtimes are valid for a relative comparison between models and cases.

steepness of the stress gradients, as illustrated by both models,
and the size of the global elements in the fillet clearly show why
there was a substantial difference in centroidal stresses between
the two models, and why there was a significant difference between the centroidal and nodal stresses in the global model. The
centroid of the global model is simply further from the surface
than the centroid of the refined model and a significant distance
from the nodes on the surface, relative to the stress gradients.

The refined model gives a good definition of the stress gradients in the fillet region that would be of use in any subsequent
crack propagation analysis. In this case, the use of the refined

Table 9.12 Stress Results of Gear Refined Model

Maximum centroidal stress				Maximum nodal stress			
σ_r	σ_a	σ_t	σ_1	σ_r	σ_a	σ_t	σ_1
Two-dimensional single tooth global model, plane strain							
506.7	228.4	254.4	588.0	904.3	382.8	371.8	1063.2
Two-dimensional filled refined model, plane strain							
868.55	341.66	270.33	1009.2	1157.22	439.48	307.71	1347.3

σ_r = radial stress
σ_a = axial stress
σ_t = tangential stress
σ_1 = maximum principal stress

model allowed for a good determination of maximum stress and stress gradients in the critical region of the gear tooth in a cost-effective manner. Using the same element density in the global model as the refined model would require 12 times the number of elements in the global model. Even by developing a transition model, using a fine gridwork in the fillet with a transition to a coarse gridwork, elsewhere, would require considerably more modeling effort and still require almost an order of magnitude more elements.

9.3.4 Example: Turbine Blade

In this example, following the three-dimensional model described in Chapter 5, Section 5.4.6, a two-dimensional plane strain, refined model is used in conjunction with the three-dimensional global model. As discussed in Chapters 4 and 5, the turbine blade is attached to the turbine rotor via a firtree root attachment design as shown in Figure 9.26. Although this attachment area is modeled with 440 (five layers of 88) three-dimensional elements, the definition of the fillets is marginal, at best, as shown in Figure 9.27.

The refined two-dimensional model uses 832 elements to model the two-dimensional blade attachment and rotor segment, as

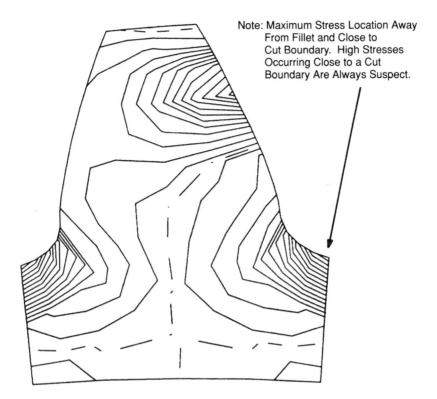

Note: Maximum Stress Location Away
From Fillet and Close to
Cut Boundary. High Stresses
Occurring Close to a Cut
Boundary Are Always Suspect.

Figure 9.24 Single gear tooth, two-dimensional global model, maximum principal stress.

shown in Figures 9.28 and 9.29. This gives almost 10 times the number of elements in the two-dimensional model as there are in one three-dimensional layer. There are 428 elements used to model the blade root and 404 elements used to model the rotor segment that has approximately the same proportions as the three-dimensional rotor segment. The use of the two-dimensional model allows gap elements to be used at the contacting surfaces. The gap elements allow for friction effects to be included in the model and for the effects of incomplete initial contact between all six load-bearing surfaces to be investigated. In other words, a situation could arise in which one or more of the

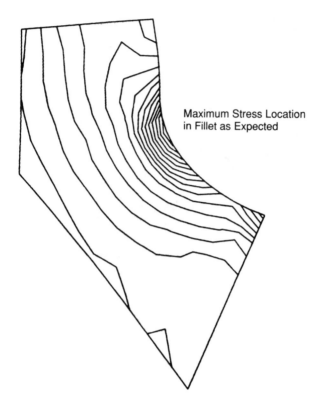

Maximum Stress Location
in Fillet as Expected

Figure 9.25 Single gear tooth, two-dimensional refined model, maximum principal stress.

"hooks" were not initially in contact. The remaining hook would be more heavily loaded and therefore more highly stressed. The use of gap elements allows for an initial open gap to be specified and included in the calculation.

The computer runtimes are given in Table 9.13. The three-dimensional global model is described in Chapter 5 and uses 735 three-dimensional solid elements. The two-dimensional, refined model includes gap elements that require an iterative solution. Substructuring was used to reduce down the problem size to only the degrees of freedom connected to the gap elements. Back substitution for stresses was carried out for the blade root only. Substructuring will be discussed in more detail in Chapter 10.

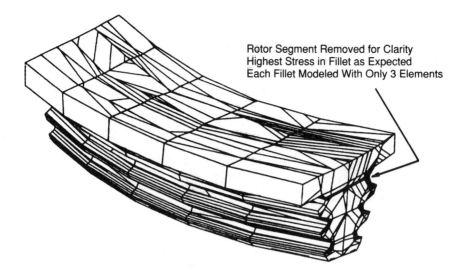

Rotor Segment Removed for Clarity
Highest Stress in Fillet as Expected
Each Fillet Modeled With Only 3 Elements

Figure 9.26 Turbine blade firtree root attachment, three-dimensional model.

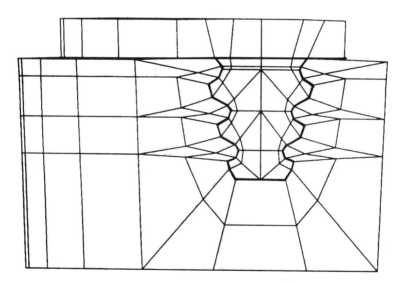

Both Blade and Rotor Segment Shown.
Each Fillet Modeled With Only 3 Elements
Along Profile by One Layer of Elements
Into Depth.

Figure 9.27 Turbine blade firtree root attachment, model profile.

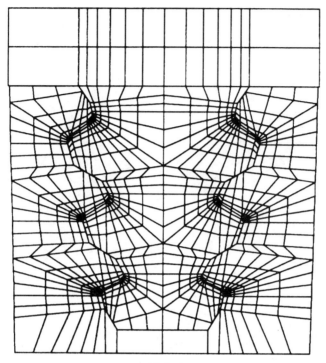

2D Refined Model has Problem of
Approximating 3D Stress Field
With 2D Plane Stress Calculation

Figure 9.28 Turbine blade root attachment, two-dimensional refined model.

The maximum principal stress occurred in the upper hook fillet
as shown in Figure 9.30. A comparison of stresses is given in
Table 9.14. The calculated elastic stress for the refined model
exceeded the yield strength of the material, indicating that there
should be some localized plastic deformation in the hook fillet. A
direct comparison of stresses is not valid here, because the global model is a full three-dimensional model, whereas the refined
model is a two-dimensional plane strain model. The main purpose
of the refined model in this case is to form a comparison of radial

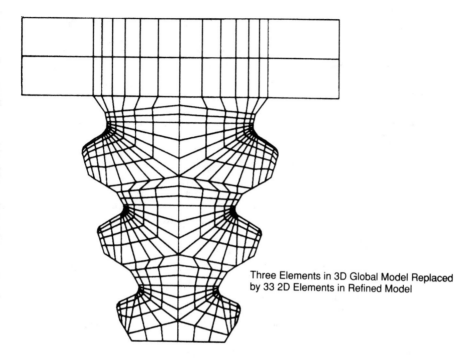

Three Elements in 3D Global Model Replaced
by 33 2D Elements in Refined Model

Figure 9.29 Turbine blade root attachment, two-dimensional refined model, blade only.

and tangential component stresses. The radial stress is the predominant stress and contributes the most to the maximum principal stress. This comparison shows that the maximum stresses
are fairly close. Both the maximum centroidal and nodal stresses for the three-dimensional global model were higher than the
corresponding two-dimensional refined stresses. It must be assumed, however, that the stress distribution into the depth of
the part should be more accurate with the refined model.

9.4 PLATE-SOLID TRANSITIONS

One particular refined modeling problem that requires consideration is the use of a two or three-dimensional refined model with

Table 9.13 Details and Computer Runtimes for Turbine Blade Refined Model

Three-dimensional turbine blade global model
  774 Elements (three-dimensional, solid)
  79 Nodes   135 active DOF
  Times (sec)

Element formulation	692.0	(0.894 sec/elem.)
Wavefront solution	2861.0	
Stress solution	1018.0	(1.315 sec/elem.)
Total	4571.0	sec

Two-dimensional refined root model
  832 Elements (two-dimensional, plane strain)
  1477 Nodes   4411 active DOF
  Times (sec)

Element formulation	126.46	(0.152 sec/elem.)
Wavefront solution	68.867	
(substructure generation)		
Gap iteration routine	45.994	
Stress solution (428 elem.)	77.710	(0.101 sec/elem.)
Total	319.03	sec

Note: Computer runtimes are valid for a relative comparison between models and cases.

plate and/or beam and solid global model. Because the plate and beam elements can model rotations and moments which the two- and three-dimensional solid elements cannot, the application of moment must be added to the refined model with care.

A typical instance in which this might occur is in a plate to solid transition where plate (or beam) elements are joined to solid elements. The suggested method of accomplishing this interface and transmitting the moment from the plate elements to the solid elements is to "bury" the plate elements one layer deep in the solid elements as shown in Figure 9.31. It is recommended that the plate element that is buried in the solid element have a reduced thickness and moment of inertia because that portion of

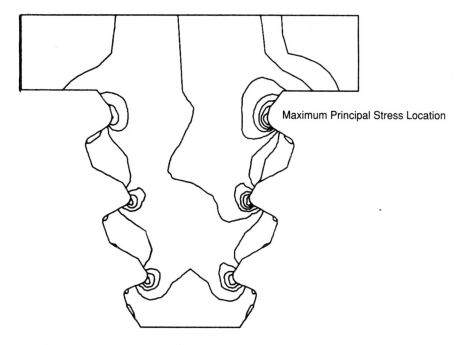

Maximum Principal Stress Location

Figure 9.30 Turbine blade, refined two-dimensional root model, maximum principal stress.

the model will be overstiff. A rule of thumb to use in reducing the stiffness is to make those elements one-tenth the thickness and stiffness of the actual plates. In this way, the degree of overstiffness will not be too severe and any numerical problems associated with large differences in stiffness within the model will be avoided.

In forming the refined model, the translation displacements from the plate elements may be applied directly to the cut boundary of the refined model. The rotation or slope of the plate nodes must also be used to adjust the displacements of the refined model node and pivot about the center of the section. This technique is illustrated in Figure 9.32.

Table 9.14 Stress Results of Turbine Blade Refined Model

	Maximum centroidal stress				Maximum nodal stress		
σ_r	σ_t	σ_a	σ_1	σ_r	σ_t	σ_a	σ_1
Three-dimensional global model, blade attachment fillet							
88,125.0	25,492.0	19,753.0	84,672.0	135,500.0	37,420.0	35,540.0	143,614.0
Two-dimensional refined model, blade attachment fillet							
80,679.0	9,294.0	26,992.0	81,298.0	102,961.0	11,072.0	34,210.0	104,759.0

σ_r = radial stress
σ_t = tangential stress
σ_a = axial stress
σ_1 = maximum principal stress

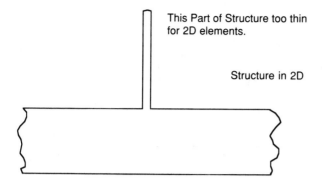

This Part of Structure too thin for 2D elements.

Structure in 2D

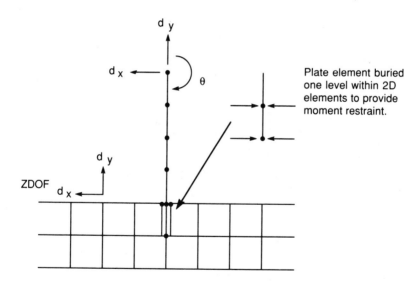

Plate element buried one level within 2D elements to provide moment restraint.

Notes: • Mass of "Buried" Beam Element should be zero.

• Stiffness of "Buried" Beam Element should be 1/10 of other Beam Elements.

Figure 9.31 Plate-to-solid transition modeling.

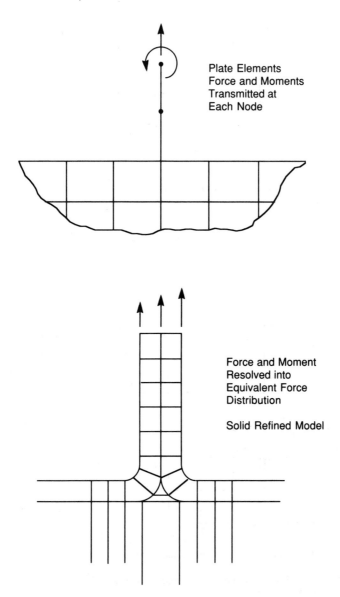

Plate Elements
Force and Moments
Transmitted at
Each Node

Force and Moment
Resolved into
Equivalent Force
Distribution

Solid Refined Model

Figure 9.32 Use of plate element displacements with solid refined models.

9.5 THERMAL EFFECTS

When thermal effects are included in the global model, it is imperative that they be transferred to the refined model. This is necessary because the displacements of the global model are a combination of thermal expansion and deformation due to applied loading. The cut boundary displacements are, therefore, also a combination of these two effects. If the refined model does not have the correct nodal temperatures applied to all nodes and is allowed to expand (or contract) to the same extent as the global model, then the refined model internal displacements will not be consistent with the cut boundary applied displacements.

An example is a global model in which thermal expansion and deformation due to applied loadings are of the same magnitude. If cut boundary displacements are applied to the boundary of the refined model and stresses are calculated at room temperature, i.e., no internal thermal expansion, then the stresses will be much higher than they actually are because the model will be "pulled" from the boundaries to the full amount of the combined displacements.

9.6 SUMMARY

Refined mesh modeling represents a cost-effective means of modeling large, complex structures with small details that can give local stress concentrations. Although the examples shown here are rather simple, actual applications to structures such as pressure vessels with valve and nozzle attachments or large machinery models with bolted joints represent opportunities to achieve better accuracy with moderate cost.

REFERENCE

1. Peterson, R. E., *Stress Concentration Factors*, Wiley, New York, 1974.

10

SUBSTRUCTURING

Many finite element models feature repetitive geometry. In problems such as gear teeth or groups of turbine blades, it may be necessary to represent a repeated component in a model. Repetitive geometry can be efficiently modeled by the use of substructuring. A substructure is a finite-element model for which the internal degrees of freedom have been solved for in terms of the boundary degrees of freedom (Figure 10.1). This reduced model can then be used as an element, even though it represents a complex geometry, and connected to other substructures or individual elements within a global model. In some finite element codes, a substructure is also known as a superelement. The terms may or may not be synonymous, depending on the system. The user must take care to consult the code documentation to determine the exact definition when those terms are used.

10.1 SUBSTRUCTURING PROCEDURES

Substructuring is performed when a repetitive geometry would require generation and solution of identical sets of element stiffness

Substructure Model Before Condensation

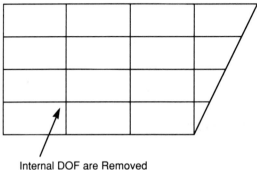

Internal DOF are Removed

Boundary DOF are Retained for Coupling with Other Substructures
or Elements

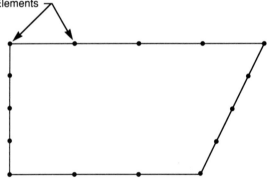

Substructure Model After Condensation

Figure 10.1 Generalized substructure.

matrices. The mathematics of substructuring are discussed in
Chapter 2, Section 2.11. In short, the individual element stiff-
nes matrices (and mass matrices, if appropriate) are assembled
as in a static solution and the matrix reduction process is begun. However, rather than completing the solution procedure
to obtain an explicit solution, the set of matrix equations is

only partially solved to provide a condensed set of equations, i.e., the substructure, and a back-substitution matrix that can provide the displacement solution once the retained or master degrees-of-freedom displacements are obtained.

In order to use substructuring, the user must understand the procedures involved. Although substructuring is aimed at maximizing the efficiency of computations, there is a price to be paid in terms of modeling and bookkeeping effort. An analysis using substructures must be performed in several steps.

1. Layout of the global model and identification of areas to be substructured. Layout and development of the substructure model. Identification of boundary nodes to be retained in the substructure.

2. Generation and condensation of the substructure stiffness matrix. This is the substructure *generation pass*.

3. Generation of the global model by using the reduced substructure as many times as required and combining it with other substructures and individual elements. The solution of the global model gives the displacements at the retained, boundary degrees of freedom of the substructures. This is the *use pass*.

4. Back-substitution of the boundary displacements into the original substructure model to obtain all displacements and stresses within the substructure. This step may be repeated as many times as necessary because when a substructure is repeated within a global model, there will be different sets of boundary displacements each time the substructure is repeated in a different location in the global model.

10.2 PRACTICAL CONSIDERATIONS

Some rules of thumb governing the use of substructuring are given in Table 10.1. Care must be taken in the application of boundary conditions in a model where substructures are used. Displacement constraints may be applied in the substructure generation pass or in the global, substructure use pass. If a node is constrained in the substructure generation pass, however, it will be constrained in all subsequent runs using the substructure. It is not possible to unconstrain a node in a subsequent substructure use pass if it has been constrained in the generation pass because the specified degrees of freedom have been removed. In the event that a particular degree of freedom is to be constrained in one "copy" of a repeated substructure and not in

Table 10.1 Rules for Substructuring

---

1. A substructure may be generated from individual elements, other substructures, or both.

2. Master nodes to be retained must be identified and specified in the input to the substructure generation run. Master nodes consist of boundary nodes that must be retained for connection with the rest of the global model and nodes for applied forces.

3. Nodal constraints specified in the substructure generation step will be constrained in all subsequent uses of the substructure and can never be released. Nodal constraints may, however, be applied in the global model to substructure master nodes.

4. Along a substructure boundary that will be used for connection to the rest of the global model, all nodes must be retained as master nodes.

5. To be cost-effective, a substructure must be used three times. If a portion of a model is repeated only once, i.e., appears twice in the structure, then substructuring will probably not be cost-effective in terms of computer runtime, not to mention the additional modeling and analysis labor time.

---

another copy, then that degree of freedom should be retained as a master in the substructure generation pass and later constrained in the use pass.

Substructuring can be a very cost-effective way of handling repetitive geometries; however, there is an overhead cost in computer time and storage. A good rule of thumb is to have a need to repeat part of a structure two times, for a total of three copies, to make substructuring cost-effective. If a part of a structure is repeated only once, i.e., appears in the structure only twice, then substructuring probably should not be used. There is a considerable amount of additional modeling and bookkeeping effort required in substructured models not required in other cases. In short, the use of substructuring is a tradeoff between computer time savings and additional labor time.

There are several other situations in which substructuring
may be used. When several different structures contain the
same component, it may be cost-effective to form a substructure
of that component and archive the substructure so that it may
be used in different models and analyses.

Another reason for substructuring is for models that are simply too large to fit into the computer as a single model. In these
events, the initial model may be broken down into a collection of
substructures that, in turn, may be reduced to their boundary
nodes, assembled together into a global model, solved for the master degrees of freedom, and the individual substructures may be
back-substituted for stresses. For a large model, substructuring
will not be as cost-effective as solving the problem as a single
large model; however, it does represent an alternative to an impossible model.

A third reason for substructuring exists when nonlinear problems are involved, such as gap elements that require an iterative
solution. In such cases, the linear portion of the model can be
generated as a substructure and reduced down to a selected set
of master degrees of freedom. The nonlinear elements can be
solved iteratively and the final stress solution may be obtained
by back-substitution.

10.3 EXAMPLE: CIRCULAR PLATE

In cases where one substructure is generated and repeated several times within a global model, the stresses within the substructure copies will not be the same but vary depending on the
location of the substructure copy within the global model. As
an example, a circular plate is shown in Figure 10.2. An unsubstructured model of the plate is shown in Figure 10.3. It
is easy to see that due to the symmetric nature of the geometry,
there is substantial repetition of the elements. However, due
to the nonsymmetric loading at the center hole, the model cannot be made smaller by using symmetry. The 10 bolt holes in
the plate for fastening require a substantial number of elements
to correctly represent the geometry and give a reasonable calculation of stress in those areas.

A substructure of one segment of the plate is shown in Figure 10.4. This substructure represents one-tenth of the plate
(36°). Because the bolt holes are evenly spaced, it is possible
to form one substructure and repeat it 10 times to form a plate
model.

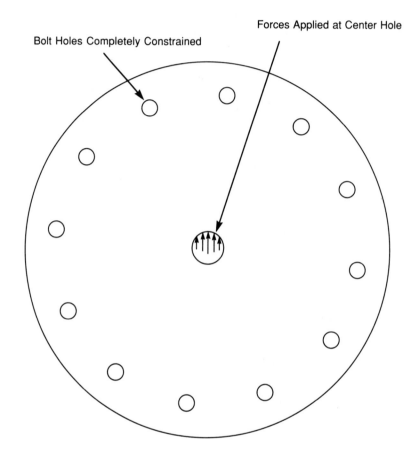

Forces Applied at Center Hole

Bolt Holes Completely Constrained

Asymmetric Pattern of Forces Makes it Impossible
to Use Symmetry to Reduce Problem Size

Figure 10.2 Circular plate substructuring example.

An unsubstructured model contains 230 elements and 260 nodes
(520 DOF) as shown in Figure 10.3. The substructure contains
23 elements and 33 nodes that are reduced down to 14 master
nodes at the symmetric boundaries. Displacement constraints

All DOF at Bolt Holes Constrained to Zero

Forces Applied to Nodes at Center Hole

230 Elements

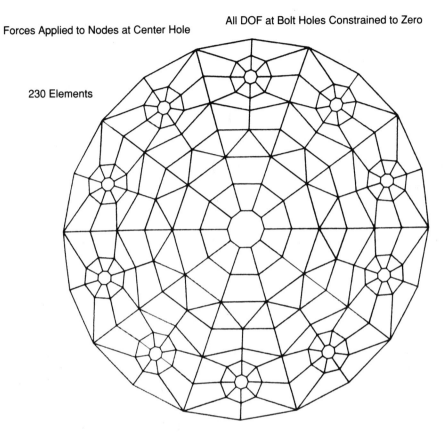

Figure 10.3 Circular plate unsubstructured model.

are applied at the bolt hole in the substructure generation run
so that those degrees of freedom are "permanently" constrained
and no additional displacement constraints need to be specified
in the subsequent use runs. The global model, consisting of 10
copies of the substructure, contains only 70 unique master nodes
(140 DOF).

Forces are applied at the center of the plate in the substructure use run at the boundary nodes so that additional nodes do
not need to be retained for the application of forces. The assembly of substructures is shown schematically in Figure 10.5.

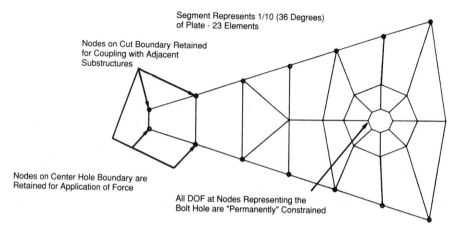

Segment Represents 1/10 (36 Degrees)
of Plate - 23 Elements

Nodes on Cut Boundary Retained
for Coupling with Adjacent
Substructures

Nodes on Center Hole Boundary are
Retained for Application of Force

All DOF at Nodes Representing the
Bolt Hole are "Permanently" Constrained

Figure 10.4 Circular plate segment substructure model.

Once the substructure has been generated and used in the static stress calculation, the first segment (0 to 36°) is selected for back-substitution to obtain the stress distribution, as shown in Figure 10.6.

A comparison of the steps involved and computer runtimes for the unsubstructured and substructured cases are given in Table 10.2. It can be seen that the total computer time required to generate the substructure was more than one-tenth of the time required to generate the elements for the unsubstructured case. However, the total computer time required for the substructured case was substantially less than for the unsubstructured case. It should also be noted that for the substructured case, only one of the segments was selected for back-substitution. There is no requirement as to how many of the segments of the global model are selected for back-substitution. All or none (if only boundary displacements are required) may be selected.

There was no substantial difference in accuracy between the substructured and nonsubstructured cases. A comparison of maximum principal stresses between the substructured and unsubstructured models is given in Table 10.3. A plot of the maximum principal stress distribution for the first 23 elements of the unsubstructured model is shown in Figure 10.7, for comparison with Figure 10.6. A stress contour plot of maximum principal stress for the complete unsubstructured model is shown in Figure 10.8.

No Internal Degrees of Freedom are Present in any of the Segments.
Boundary Nodes Between Segments Have Been Retained.
Nodes and DOF Representing Bolt Holes Have Been Constrained
in Substructure Generation Run and are no Longer Active.

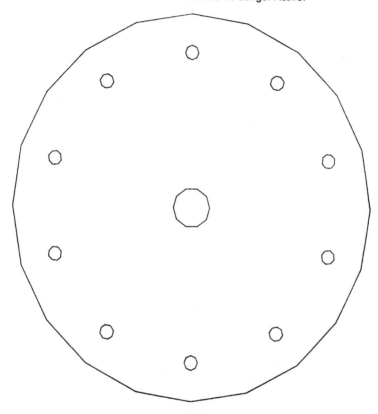

Figure 10.5 Circular plate assembly of substructures.

10.4 EXAMPLE: GEAR SEGMENT

The gear segment used to illustrate basic modeling principles in
Chapters 4 and 5 is an excellent, practical example of the use of
substructuring. The two-dimensional three-tooth model discussed in Chapter 5 represents the unsubstructured case; however, the node and element pattern for each of the teeth was

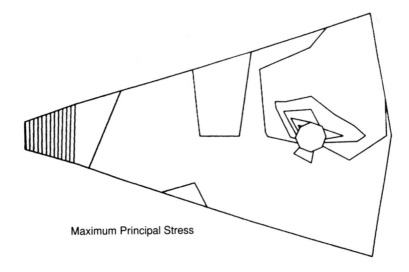

Maximum Principal Stress

Figure 10.6 Circular plate stress results: First segment of sub-structured model.

identical. Due to the unequal loading between the teeth, which is a function of the gear's instantaneous angular position relative to its mating gear, the stress pattern in all three teeth will not be the same. In addition, the bending of one tooth causes a compressive stress in the fillet of the next tooth that cannot be modeled with a single tooth model. For these reasons, the three-tooth model gives a more accurate representation of the fillet bending stresses than does the single tooth model.

The generation of three sets of identical stiffness matrices and their solution is, however, inefficient. The problem can be solved to give, for all practical purposes, identical results with a savings in computer runtime by the use of substructuring. Figure 10.9 shows the unsubstructured model from Chapter 5 with 180 elements. Figure 10.10 shows the single-tooth substructure with 60 elements and indicating those nodes that will be retained in the substructure. The nodes to be retained are those that are necessary for joining the substructures together along their cut boundaries and those nodes necessary for applying forces. It must be noted that all of the nodes on the profile

Table 10.2 Comparison of Computer Runtimes for Substructured and Unsubstructured Circular Plate Example

Structured		Unsubstructured	
Generation pass			
Element generation	23 elements	Element generation	230 elements
	3.726		37.079
	(0.162 sec/elem.)		(0.161 sec/elem.)
Substructure condensation	2.257	Displacement solution	27.730
Misc.	8.300	Element stresses	19.558
		Element forces	11.988
	14.283 sec	Misc.	36.866
			133.221 sec
Use pass	10 Copies of		
	substructure		
Read-in data	27.591		
Displacements	2.870		
Write files	4.912		
	35.403 sec		
Stress pass	23 elements		
Back-substitution	1.885		
Element stresses	1.472		
	(0.064 sec/elem.)		
Nodal forces	0.920		
Misc.	1.370		
	5.647 sec		
Total time	55.333 sec		

Note: Computer runtimes are valid for a relative comparison between models and cases.

Table 10.3 Stress Results of Circular Plate: Substructured and Nonsubstructured Models

	Maximum centroidal stress			
	σ_x	σ_y	τ_{xy}	σ_1
Substructured model, first segment	-124.44	47.23	-19.33	50.001
Unsubstructured model, first segment	-124.04	47.64	-19.78	49.889

σ_x = X direction (horizontal axis) stress
σ_y = Y direction (vertical axis) stress
τ_{xy} = shear stress
σ_1 = maximum principal stress

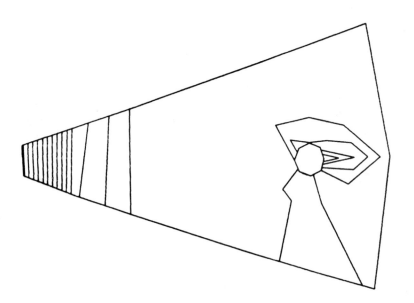

Figure 10.7 Circular plate unsubstructured case, first 23 elements.

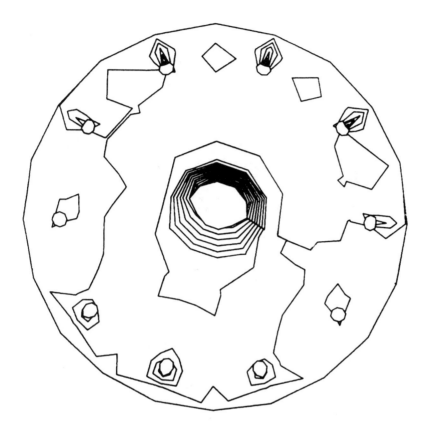

Figure 10.8 Circular plate unsubstructured case: Complete plate.

to which forces will be applied are required on every tooth; however, it is necessary to retain any node that will be required later as a location for applied forces. Figure 10.11 shows the condensed substructure schematically and the assembly of substructures. Figure 10.11 does depict the gear tooth divided into elements although at this point, after the condensation process, the internal degrees of freedom have been removed. The resulting stiffness matrix represents the relative stiffness between the remaining boundary degrees of freedom. In effect, the resulting substructure is a gear tooth "element."

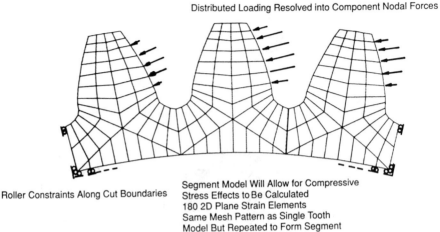

Distributed Loading Resolved into Component Nodal Forces

Roller Constraints Along Cut Boundaries

Segment Model Will Allow for Compressive
Stress Effects to Be Calculated
180 2D Plane Strain Elements
Same Mesh Pattern as Single Tooth
Model But Repeated to Form Segment

Figure 10.9 Three-tooth gear segment: Two-dimensional model.

Table 10.4 gives a comparison of the computer runtimes for
the substructured and unsubstructured three-tooth models.
Table 10.4 shows that for the three-tooth model, which by the
rule of thumb in Table 10.1 represents the minimum number of
repeated structures necessary for a breakeven point, the substructured model required 63% of the total computer runtime of
the unsubstructured case. The substructured case, it must be
noted, only obtained stresses for one of the three gear teeth.
This is, however, a practical situation because it may not be
necessary to calculate stress in all the teeth. It should be obvious from the loading pattern which tooth is going to be the
most highly stressed. If it is not intuitively obvious, then an
examination of the displacements from the use pass may give
some direction. In any event, even if stresses were calculated
in all three teeth, then an estimate for the total runtime for the
substructured case would be about 91% of the total time for the
unsubstructured case.

A closer examination of the computer runtimes in Table 10.4
points out, for the substructured case, the relative computer
time for the various phases. A linear extrapolation of times for
the use pass could be used to obtain ball-park time estimates for

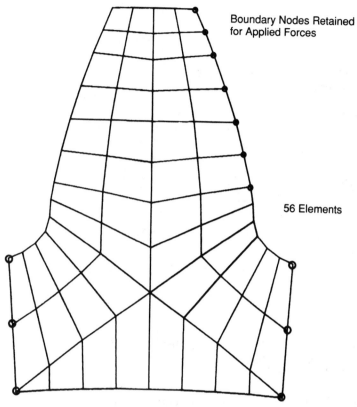

Boundary Nodes Retained
for Applied Forces

56 Elements

Boundary Nodes Retained
for Coupling

Boundary Nodes Retained
for Coupling

Figure 10.10 Single gear tooth two-dimensional substructure
model.

other numbers of gear teeth in the segment. A comparison of
the relative and total runtimes shows why three repeated structures are necessary to make it cost-effective to use substructuring. Stresses for the substructured and unsubstructured
cases were virtually identical.

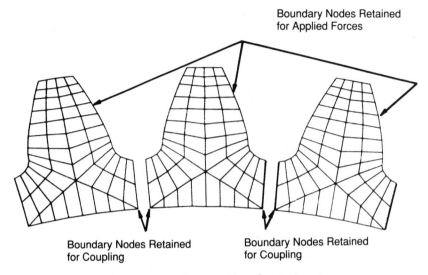

Boundary Nodes Retained
for Applied Forces

Boundary Nodes Retained
for Coupling

Boundary Nodes Retained
for Coupling

Figure 10.11   Gear segment assembly of substructures.

10.5   TURBINE BLADE ATTACHMENT: REFINED MODEL

The turbine blade refined model from Chapter 9, Section 9.3.4 is an example of the use of substructuring to cut down on the computer time required for a nonlinear calculation. This two-dimensional plane strain model is shown in Figure 10.12. The model consists of a blade root section with 428 elements, a segment of the mating turbine rotor with 405 elements, and 18 two-dimensional gap elements to represent the contact between the blade and rotor load-bearing surfaces. The use of the gap elements makes the problem geometrically nonlinear and the gap elements require an iterative solution. The gap elements are allowed to be open or closed and to slide sideways. Equilibrium is achieved when all gap elements find their position, either open or closed, and there is no further sliding. This typically requires between 5 and 10 iterations for this type of model. In this case, substructuring is used to reduce down the initial models to only those nodes that define the gap elements and those

Table 10.4 Comparison of Computer Runtimes for Substructured and Unsubstructured Gear Segment Example

Substructured			Unsubstructured		
Generation pass			Element generation	180 elements	
Element generation	60 elements			28.800	(0.160 sec/elem.)
	9.660	(0.161 sec/elem.)	Displacement solution	21.702	
Substructure condensation	5.888		Element stresses	15.306	
Misc.	21.652		Element forces	9.382	
	37.200 sec		Misc.	28.852	
				104.042 sec	
Use pass	3 Copies of substructure				
Read-in data	9.591				
Displacements	0.970				
Write files	2.672				
	13.203 sec				
Stress pass	60 elements				
Back-substitution	4.917				
Element stresses	3.840	(0.064 sec/elem.)			
Nodal forces	2.402				
Misc.	3.574				
	14.733 sec				
Total time	65.136 sec				

Note: Computer runtimes are valid for a relative comparison between models and cases.

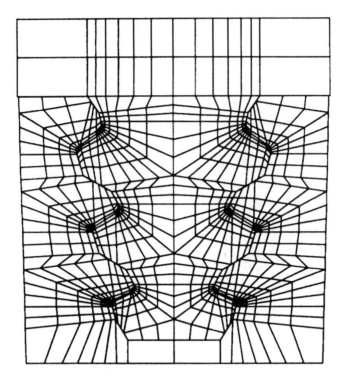

Figure 10.12 Turbine blade attachment: Refined two-dimensional model.

nodes on the blade root to which forces are applied. Nodes to which displacement constraints are applied on the rotor segment are constrained in the generation pass.

Figure 10.13 shows schematically the final blade root and rotor segment substructures assembled together with the gap elements. This example case differs from the previous two in that this case combines two different substructures together with conventional elements in the use pass. Figure 10.14 shows the resulting two-dimensional maximum principal stress distribution in the blade attachment portion.

Table 10.5 shows the breakdown of computer runtime for the entire analysis. It shows that the time for the gap element iterations was only 32 sec for five iterations, a little over 6 sec per iteration. This should be contrasted with the time for the

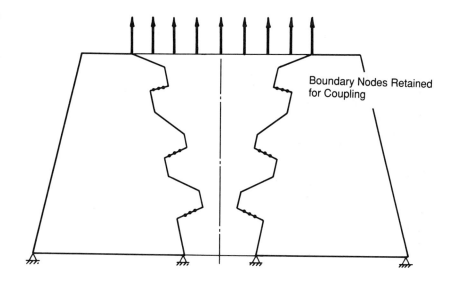

Boundary Nodes Retained
for Coupling

⟋ Constraints Applied to Boundary of Assembled Substructures

↑ Forces Applied Boundary of Assembled Substructures

Figure 10.13 Turbine blade attachment: Assembly of substructures.

substructure condensation, 461.02 sec, which would be roughly the same as the time to solve for all of the elements at each iteration if substructuring were not used.

Chapter 9, Section 9.3.4 gives the remaining details of stresses and comparison with the global three-dimensional model. The purpose of the discussion here is to point out another example of substructuring to minimize the computer runtime for a nonlinear problem.

10.6 SUMMARY

Substructuring does not change the overall accuracy of the analysis, nor does it represent an approximation or require any

Table 10.5 Computer Runtimes for Substructured Two-Dimensional Refined Turbine Blade Example

	Substructure 1	Substructure 2
Generation pass		
Element generation	428 elements 64.891 (0.152 sec/elem.)	
Substructure condensation	41.622	
Misc.	155.342 ————— 261.86 sec	
Element generation		405 elements 53.815 (0.133 sec/elem.)
Substructure condensation		29.130
Misc.		116.21 ————— 119.16 sec
Use pass		
Read-in data	27.65	
Gap iterations	32.56 (18 gap elements, 6 iterations) ————— 60.209 sec	
Stress pass		
Back-substitution	428 elements 21.936	
Element stresses	55.040 (0.129 sec/elem.)	
Misc.	0.227 ————— 77.203 sec	

Note: Computer runtimes are valid for a relative comparison between models and cases.

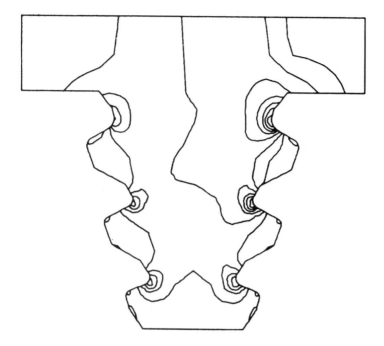

Figure 10.14 Turbine blade attachment: Two-dimensional stress distribution.

additional modeling assumptions. It is merely a method of reducing computer runtime for repetitious structures, complex
structures, or structures with limited nonlinear effects. The
drawback to substructuring is that it requires more of the an-
alyst's time in setting up the models and in bookkeeping during the calculation process. Substructuring is, at a minimum,
a three-step procedure even for a linear static analysis. Files
must be kept track of and archived. The opportunitites for
using substructuring are usually fairly obvious and the reduction in computer time can make the additional effort well worthwhile.

11

DYNAMIC FINITE ELEMENT MODELING

11.1 OVERVIEW OF DYNAMIC MODELING CONSIDERATIONS

Dynamic finite-element analysis constitutes a broad subset of the finite element method. Dynamic finite element analysis is commonly used on even simple structures for which a textbook solution does not exist. For linear static problems and even thermal problems, there are exact or empirical solutions available for simple structures through sources such as Roark's handbook for linear static [1]. In the case of vibrations, solutions for natural frequencies and mode shapes exist for only the most simple beam- and plate-type structures in sources such as Timoshenko's *Vibration Problems in Engineering* [2] or *The Shock and Vibration Handbook* [3].

Nonfinite element forced-response calculations are generally restricted to discrete spring-mass systems using techniques such as the transfer matrix approach. These calculations require that the structure be discretized by hand and the problem size be

limited. Forced-response calculations for continuous structures
such as beams and plates become difficult because of the addition
of damping and time-dependent, distributed loading. In most
cases, these forced-response calculations are approximated by
modal superposition techniques.

Dynamic finite-element analyses can be roughly grouped into
three main categories:

1. Modal analysis that calculates free vibration natural frequencies and mode shapes.
2. Harmonic analysis that calculates the forced response of
   a structure to a sinusoidal forcing.
3. Transient analysis that calculates the forced response of
   a structure to a nonharmonic, time-varying loading such
   as an impact, step, or ramp forcing.

Real-life dynamic problems can become complex due to nonlineari-
ties. A simple bouncing ball is an example of a nonlinear transient dynamic problem. In military applications, for example,
plasticity and large displacements may be combined with transient dynamic response to model the response of a bomb hitting its
target.

11.2 MATHEMATICAL FUNDAMENTALS

The mathematics of finite element dynamics are covered in Chapter 2, Section 2.10. A brief overview is given again to relate
the fundamentals to the appropriate modeling considerations. In
forming the matrix equations of motion for dynamic finite element
analysis, it is necessary to form element mass matrices, as well
as element stiffness matrices, and assemble these together to
form a global mass matrix. The mass matrix represents the distributed mass of the structure resolved to the nodal points in a
method analogous to the resolution of the structural stiffness to
the stiffness between nodal points. In forming a consistent element mass matrix, the element density is integrated over the element volume.

Dynamic forced response is a function of three parameters:
(1) structural stiffness and mass properties, represented by
the finite element analysis model; (2) dynamic forces including
amplitude, frequency, and relative phase angle; and (3) damping.

11.2.1 Damping

Damping is a result of one or more of three phenomena: material damping due to internal friction of the structure (also known as hysteretic damping), aerodynamic or viscous damping that is due to the motion of the structure vibrating in air or in a liquid, and friction or coulomb damping due to sliding friction between discrete parts of a structure. Damping effects are input either as a separate damping matrix or as multipliers to be applied to the stiffness and mass matrices.

Numerical damping occurs in finite element dynamic forced-response calculations in addition to any specified damping. The amount of numerical damping depends on the solution method employed and the length of the time step used. Figure 11.1 shows a plot of numerical damping vs. time step size for the Houbolt

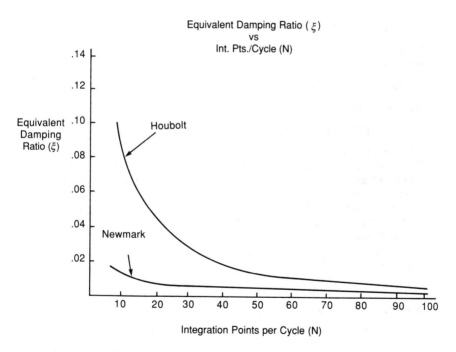

Figure 11.1 Numerical damping vs. time step size.

and Newmark methods of transient-response analysis [4]. Numerical damping represents a net damping ratio. It is advisable to calculate or at least estimate the amount of numerical damping that will occur in a particular calculation. If it is greater than the sum of the actual damping in the structure, then the time step size should be adjusted to give the appropriate damping ratio and no additional damping will need to be specified.

Damping has the greatest effect on the response of a structure when it is being excited near resonance. Figure 11.2 shows the dynamic multiplier, i.e., the ratio of dynamic response to static response, as a function of frequency and damping ratio. This figure shows that away from resonance, the actual value of the damping ratio is not as significant as it is at or near resonance.

11.2.2 Equations of Motion

For linear static calculations, the matrix displacement equations are

$$[K] \ \{x\} = \{F\}$$

To calculate undamped natural frequencies and mode shapes, a mass matrix is added

$$[M] \ \{\ddot{x}\} + [K] \ \{x\} = 0$$

Assuming sinusoidal motion gives

$$\{-\omega^2 \ [M] + [K]\} \ \{X_0\} \sin(\omega t + \phi) = 0$$

This equation represents an eigenvalue problem that can be solved for a set of eigenvalues (natural frequencies) ω and a unique eigenvector (mode shape) $\{x\}$ for each eigenvalue. There are a number of standardized numerical techniques used for the solution of the eigenvalue problem that include the Jacobbi, Householder, Givens, modified Givens, inverse power sweep, and subspace iteration.

For forced response, a damping matrix and a forcing vector are added

$$[M]\{\ddot{x}\} + [C]\{\dot{x}\} + [K]\{x\} = \{F(t\omega + \phi)\}$$

Again, if we assume sinusoidal motion,

$$\{-\omega^2 \ [M] + j \ \omega \ [C] + [K]\} \ \{X_0\} \sin(\omega t + \phi) = \{F(\omega t + \phi)\}$$

At Resonance Vibration Amplitude is Totally a Function of Damping

Dynamic Amplification Factor

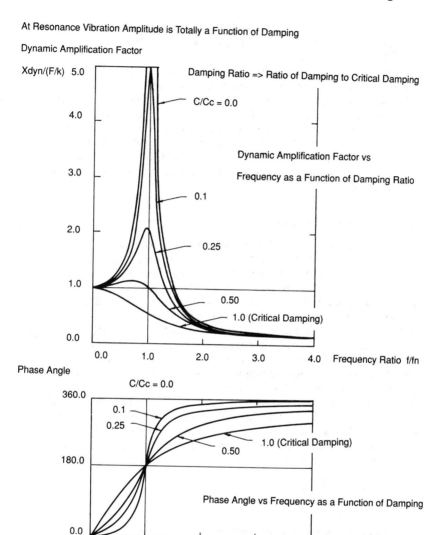

Figure 11.2 Dynamic response near resonance as a function of damping.

Because the damping term is 90° out of phase with the stiffness and mass terms, it must be handled with complex arithmetic (note $j = \sqrt{-1}$).

The forcing vector may consist of sets of either harmonic forcing, one set for each frequency, or sets of time-varying transient forcing, one set for each time increment. The forced-response calculations are also performed by a number of well-known time-integration techniques such as the Newmark beta, Houbolt, Wilson, and central difference methods. Most of the commercial finite element codes use more than one method, depending on whether the problem is harmonic or transient, linear or nonlinear.

11.2.3 Dynamic Condensation

Dynamics problems require the resolving of the matrix equations for multiple frequencies or multiple time steps, making them substantially more computer-intensive than simple linear static problems. For that reason, dynamic condensation techniques exist for the reduction of the global problem to a reduced set of dynamic degrees of freedom that are deemed sufficient to characterize the dynamic response of the structure. A popular form of dynamic condensation is Guyan reduction [5], which resembles substructuring. With Guyan reduction, the original stiffness and mass matrices are reduced to a specified number of dynamic degrees of freedom. The stiffness terms are reduced independently and the mass terms are then redistributed according to the reduced stiffness matrix with more mass assigned to rigid regions of the structure. For example, if node number 6 is to be condensed out and its mass reassigned to nodes 5 and 7, the equivalent stiffness of k'_{5-7} is simply

$$k'_{5-7} = \frac{1}{\dfrac{1}{k_{5-6}} + \dfrac{1}{k_{6-7}}}$$

The mass of node 6 will be distributed to nodes 5 and 7 according to the relative stiffness of k_{5-6} and k_{6-7}.

$$m'_5 = m_5 + m_6 \frac{k_{5-6}}{k'_{5-7}}$$

Dynamic degrees of freedom are generally distributed throughout the structure. For cases in which a true dynamic substructure is to be created, dynamic degrees of freedom are initially retained along the boundary to connect the substructures and removed in subsequent Guyan reduction runs.

11.3 DYNAMIC FINITE ELEMENT MODELING CONSIDERATIONS

Dynamic finite-element modeling requires more knowledge of the fundamentals of vibrations than the knowledge of specific finite-element procedures. A working knowledge of the structure to be analyzed is very important due to the complexity of the input required: dynamic forces that include amplitude, frequency, and phase angle damping that may need to be approximated as a single parameter and the selection of dynamic degrees of freedom from the original model. Each of the input items requires certain assumptions to be made that are more involved than those for linear static analysis.

11.3.1 Global Model

Dynamic finite element analysis requires, in effect, that two models be developed: a global model and a reduced model. The global model must contain sufficient detail to characterize the stiffness and mass properties of the structure. The amount of detail to include in the global model will depend on whether or not dynamic stresses are of concern and at what location. In cases where dynamic displacement response is the only required output, it may be possible to get away with a more coarse model. For example, if a cantilever beam joins a wall and there is a contoured fillet at the intersection, then the bending stresses in the beam at the wall will be influenced by the shape of the fillet. If the dynamic response or mode shapes are the only interest, then whether the fillet is represented in the model or not will have an insignificant bearing on the results. See Figure 11.3.

A coarse model will do a better job of predicting natural frequencies than predicting mode shapes. The higher the modes of interest, the more detailed the model needs to be. This applies to forced-harmonic and transient-response calculations as well as modal calculations.

Symmetry cannot be used as effectively for dynamic models as it is for static models. In most structures, there are asymmetric modes that may be missed if symmetry is assumed. The use of symmetry virtually requires the details of the mode shapes to

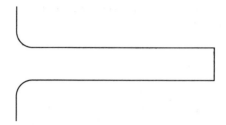

Stress Model Requires Good Detail in the Highly Stressed Area

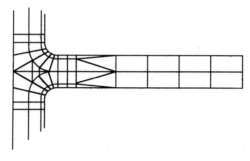

Dynamic Models Require Less Detail to Calculate Dynamic Response - Must Give Representation of Structural Stiffness and Mass.

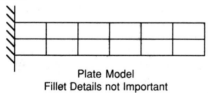

Plate Model
Fillet Details not Important

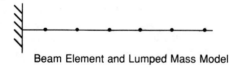

Beam Element and Lumped Mass Model

Figure 11.3 Simple beam in bending: Comparison of models.

be known in advance to prevent missing a mode(s). In static analysis, a portion of the structure may be selected for modeling with appropriate boundary conditions applied. In dynamic modeling, it may not be possible to select only a portion of a structure for inclusion into the model. Therefore, many dynamic models tend to include more of a structure with a coarser gridwork than a corresponding static model.

Plate and beam elements work well for global models in which only dynamic displacement response is required. For plate- and beam-type structures, these elements give good dynamic results with only a few elements. Stresses at critical locations such as welded joints will only be possible if a second refined submodel is used.

11.3.2 Reduced Model

The reduced model consists of the set of dynamic degrees of freedom, a subset of the global degrees of freedom. The dynamic degrees of freedom are specified as a node number and one or more of the degrees of freedom for that node. For a specified node, not all degrees of freedom need to be identified as dynamic degrees of freedom.

Because the actual calculations are performed with the reduced model, the selection of the set of dynamic degrees of freedom is often more important than the gridwork of the global model. General rules of thumb for the selection of dynamic degrees of freedom are given in Table 11.1.

All three types of analysis—modal, harmonic, and transient—may be performed with a reduced model. When more than one type of analysis is to be performed, the same reduced model may be used, thereby only requiring the dynamic condensation step to be performed once.

11.4 MODELING CONSIDERATIONS FOR THE THREE TYPES OF ANALYSIS

In modeling a dynamics problem, the objectives of the analysis will help to determine what types of analyses are required. If only natural frequencies and mode shapes are required, then obviously only a modal analysis is required.

If the forced response of the structure is required, then it must be determined whether or not the dynamic force is periodic

Table 11.1 Guidelines for Selection of Dynamic Degrees of Freedom (DDOF)

- The number of DDOF should be at least twice the highest mode of interest.

- For modal analysis the number of reduced modes will be equal to the number of DDOF so that only the "bottom half" of the calculated modes should be considered to be accurate.

- DDOF should be placed in areas of large mass and rigidity. These areas tend to drive the mode shape.

- DDOF should be distributed in such a manner to describe the anticipated mode shapes.

- A dynamic degree of freedom must be selected at each point of dynamic force application.

- DDOF must be used with gap elements.

- DDOF do not have to be selected constraint points as long as the constraints are permanent. These constraints may be specified in the global model.

- The total computer runtime is a weak function of the number of DDOF. The total computer runtime is not severely affected unless an exorbitant number of DDOF are used. If the number of DDOF is 1-2% of the total, original degrees of freedom in the model then the addition of a few more or less DDOF will make an insignificant change in the total computer time.

- For plate type structures, DDOF in the out-of-plane direction should be emphasized. In plane, DDOF will be less effective and should not be used unless necessary. In addition, rotational degrees of freedom are not necessary.

---

or transient. In the case of periodic force, the force must be broken down into its harmonic components by Fourier analysis in order to perform a forced harmonic calculation. If the force cannot be decomposed into its harmonic components or it is transient, then the response must be calculated by a time-transient calculation. Any force-time history may be input to a transient analysis, even a sinusoidal force, although harmonic-response calculations are much more efficient.

In either case, it is recommended that a modal analysis be performed first to identify the natural frequencies and mode shapes in order to give insight into the forced-response results. If possible, the natural frequencies and mode shapes should be verified prior to continuing with a forced-response analysis. This may be accomplished by experimental modal analysis or a simple rap test using an accelerometer and spectrum analyzer.

11.4.1 Modal Analysis

Modal analysis is the calculation of free vibration natural frequencies and mode shapes. Every structure has natural frequencies of vibration and a unique mode shape associated with each frequency. For a discrete spring-mass system, the number of natural frequencies and mode shapes equals the number of degrees of freedom. For a single degree of freedom system, the natural frequency is given by $f = 2\pi \sqrt{k/m}$, hz where m is the mass and k the spring stiffness. A continuous structure has theoretically an infinite number of natural frequencies, although in practice only the first few modes are considered. The lowest modes are generally the easiest to excite. Higher modes with their more complex mode shapes require more energy to excite and therefore are of less concern. Mode shapes of a simple beam with various constraint conditions are shown in Figure 11.4.

Natural frequencies are of importance to the engineer because if a natural frequency occurs near an exciting frequency in a machine, such as a harmonic of the running speed, a resonant condition may occur that could result in high vibration and damage to the machine. In any vibration analysis, whether it be for a new design or an existing vibration problem, the identification of the system's natural frequencies and mode shapes is the first step. Most design engineers avoid vibration problems simply by making sure that there are no natural frequencies coincident with potential sources of excitation in the machine.

With finite element analysis, the continuous structure is initially discretized into a finite number of degrees of freedom. In many cases, the model may be further reduced by the use of Guyan reduction that condenses the model into a small number of dynamic degrees of freedom. The reduced model is solved by a direct procedure such as the Jacobbi or Householder methods to give a number of natural frequencies and reduced mode shapes equal to the number of reduced dynamic degrees of freedom. A good rule of thumb, as pointed out in Table 11.1, is to allow twice

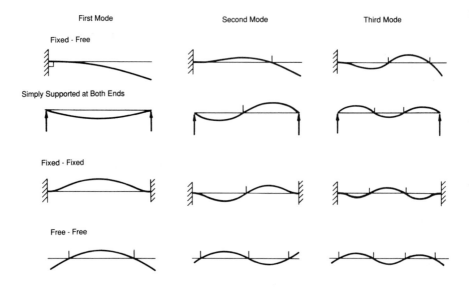

Figure 11.4 Mode shapes of a simple beam.

as many dynamic degrees of freedom as modes of interest. The selection of dynamic degrees of freedom is, in many cases, more significant in determining the accuracy of the calculated natural frequencies than the global model. Dynamic degrees of freedom for a simple beam are shown in Figure 11.5.

An alternative method of solving for natural frequencies is subspace iteration that does not require the problem size to be reduced, but rather begins by solving for the lowest modes first and can be specified to stop at any number of modes.

Following the calculations of natural frequencies with the reduced model, a back-substitution step may be carried out to obtain the complete mode shape of the structure. In this step, the displacements of the dynamic degrees of freedom are used along with the back-substitution matrix generated during the dynamic condensation step to obtain the displacements of all the degrees of freedom in the model. At this time, the modal stresses may be obtained throughout the structure; however, care must be taken to adjust the modal displacements to actual displacements.

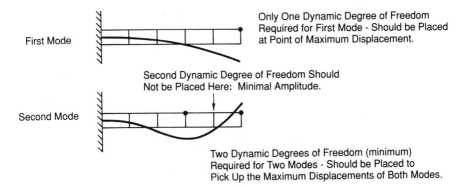

First Mode

Only One Dynamic Degree of Freedom
Required for First Mode - Should be Placed
at Point of Maximum Displacement.

Second Dynamic Degree of Freedom Should
Not be Placed Here: Minimal Amplitude.

Second Mode

Two Dynamic Degrees of Freedom (minimum)
Required for Two Modes - Should be Placed to
Pick Up the Maximum Displacements of Both Modes.

Figure 11.5 Selection of dynamic degrees of freedom for a simple beam.

11.4.2 Simple Plate Modal Analysis Example

A simple example of finite element modal analysis is the calculation of the first bending modes of a cantilevered flat plate (Figure 11.6). Dimensions of the plate are 1 × 2 × 0.1, Young's
modulus is 30,000,000, Poisson's ratio is 0.30, and the density
is 0.00073. Along the mounting edge, all degrees of freedom are
constrained. The plate is modeled with two models, the first with
32 quadrilateral plate elements with six degrees of freedom at each
node and the second with 32 three-dimensional solid elements
(LDADS) with three degrees of freedom at each node. A Guyan
reduction procedure was employed to reduce the problem to selected sets of dynamic degrees of freedom normal to the plate.
The Householder method was used to solve for the natural frequencies and reduced mode shapes. Full subspace iteration was
also used to calculate the natural frequencies and mode shapes.

The number of dynamic degrees of freedom was varied for both
models, and the results (Tables 11.2 and 11.3) indicate that the
first mode converged with the fewest DDOF and the higher modes
required more DDOF due to the increased complexity of the mode
shapes. Mode shape plots for the plate element model are shown
in Figures 11.7—11.11. When only three dynamic degrees of freedom (DDOF) were used, only the first and third modes were

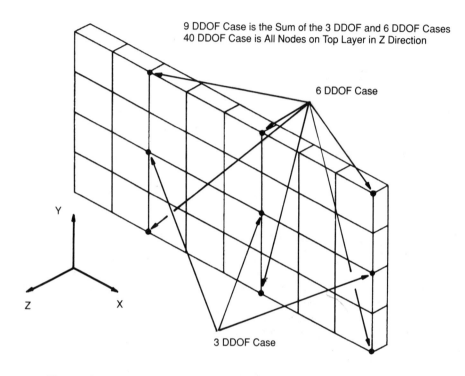

Figure 11.6 Flat plate modal analysis example.

picked up. The pattern of DDOF for the various cases is shown in Figure 11.6. For the 3, 6, 9, and 40 DDOF cases, the DDOF were specified as one per node and only the out-of-plane component was used.

A comparison of the patterns of DDOF with the second, twisting mode shape makes it obvious why the 3 DDOF model with the DDOF lying on the centerline of the plate could not calculate a second mode. The fourth mode is an in-plane bending mode so that only the cases that carried in-plane DDOF (27, 45, and subspace iteration), as shown in Table 11.3, were capable of picking

Table 11.2 Modal Analysis Results of Solid and Plate Elements with Variable Numbers of DDOF—Cantilevered Plate Model

Dynamic degrees of freedom		Frequencies, hz				
		MODE 1	MODE 2	MODE 3	MODE 4	MODE 5
3	Plate	840.81	*	5487.15	*	
	Solid	841.21	*	5618.20	*	
6	Plate	841.98	3664.27	5604.35	*	12,866.12
	Solid	842.11	3608.07	5709.09	*	12,736.15
9	Plate	839.68	3664.27	5336.87	*	12,866.12
	Solid	839.98	3608.07	5441.12	*	12,736.15
40	Plate	838.44	3652.08	5292.62	*	12,378.20
	Solid	839.61	3587.64	5367.49	*	11,975.81
Full subspace	Plate	839.20	3637.38	5245.70	7063.34	12,874.89
iteration	Solid	839.61	3587.58	5367.32	7127.60	12,537.55
Ritz method		846.00	3638.00	5266.00		11,870.00

*Mode not calculated due to an insufficient number of dynamic degrees of freedom to pick up the mode shape.
Notes: All selected DDOF are out-of-plane. Mode 4 is an in-plane mode which cannot be picked up by the out-of-plane DDOF.

up that mode. This points out that one of the "dangers" involved in selecting DDOF is that a finite element program will not give an error message if a mode is missed. The results look perfectly good for all the cases with no warning that modes may have been entirely missed.

The computer runtimes are shown in Tables 11.4 and 11.5. These numbers indicate that the time for the actual eigenvalue (natural frequency) solution is a small fraction of the total runtime when the Guyan reduction procedure was used first. The results of Table 11.5 show that the time for Guyan reduction increases with the number of DDOF, but it is not a directly proportional increase. The eigenvalue solution time is a stronger

Table 11.3 Modal Analysis Results of Plate Elements with Different Patterns of DDOF—Cantilevered Plate Model

Dynamic degree of freedom pattern	Frequencies, hz				
	MODE1	MODE2	MODE3	MODE4	MODE5
3 - Out of plane	840.81	*	5487.15	*	
6 - Out of plane	841.98	3664.27	5604.35	*	12,866.12
9 - Out of plane	839.68	3664.27	5336.87	*	12,866.12
27 - 9 nodes, all translations - out of plane and in plane	839.68	3664.27	5336.88	7091.22	12,945.95
45 - 9 nodes, all translations and two out of plane rotations	839.44	3652.08	5292.62	7091.22	12,924.15
40 - Out of plane	838.44	3652.08	5292.62	*	12,378.20
Full subspace iteration	839.20	3637.38	5245.70	7063.34	12,874.89
Ritz method	846.00	3638.00	5266.00		11,870.00

*Mode not calculated due to an insufficient number of dynamic degrees of freedom to pick up the mode shape.
Note: Mode 4 is an in-plane mode which cannot be picked up by the out-of-plane DDOF.

function of the number of DDOF than is the time for Guyan reduction. When the total runtime is considered as a function of the number of DDOF, it can be seen that the number of DDOF influences the total runtime, but that for the benefit of increased accuracy, it is foolish to skimp on the DDOF.

The first five natural frequencies were also calculated using the subspace iteration method that uses all degrees of freedom. The results were almost the same as the 45 dynamic degree-of-freedom case; however, the subspace iteration calculation required twice as much computer time. The full subspace iteration

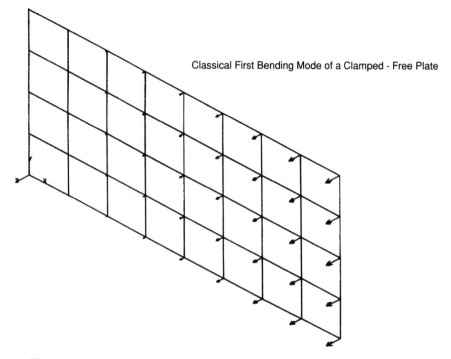

Classical First Bending Mode of a Clamped - Free Plate

Figure 11.7 Flat plate example: Mode 1.

technique does not require the identification of dynamic degrees
of freedom by the user. The results of the calculations with various dynamic degrees of freedom are given in Tables 11.2 and 11.3.

Eight-node solid elements were substituted for plate elements
and showed very little difference in results compared with the
plate-element model. A comparison of the results is shown in
Table 11.2. A comparison of computer runtimes is given in Table
11.4. The time for the Guyan reduction/Householder solution
showed that the solid elements took about 16% more total computer
time than the plate elements. This difference can be traced directly to the difference in the total degrees of freedom (240:200).
The time for element formulation was virtually identical. For the
subspace iteration, the time required for the plate-element model
was about 6% more than that for the solid element model.

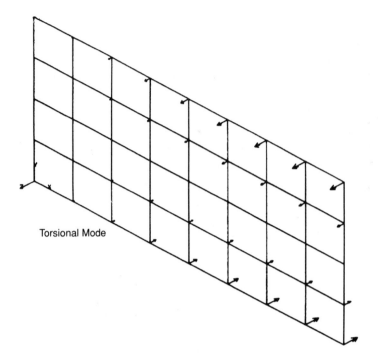

Torsional Mode

Figure 11.8 Flat plate example: Mode 2.

This example shows how the improper selection or use of too few dynamic degrees of freedom can cause erroneous results and modes to be missed entirely. The example also shows the economy in intelligently selecting the dynamic degrees of freedom and working with a reduced model. This example is similar to that presented by Zienkiewicz in [6], Chapter 17.

11.4.3 Steam Turbine Blade Modal Analysis Example

A model of a single, freestanding steam turbine blade, similar to that discussed in Chapters 4 and 5, is used to illustrate the sensitivity of natural frequencies to element density and dynamic degree of freedom selection and density [7]. This turbine blade,

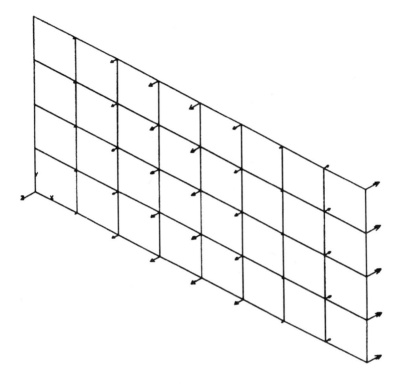

Figure 11.9 Flat plate example: Mode 3.

shown in Figure 11.12, is slightly different than the blade discussed previously in that it does not have the cover band segment.

Two parametric studies were performed: (1) the number of
elements along the length was varied as 6, 12, 18, 24, 30 and
(2) the number of dynamic degrees of freedom was varied as 12,
24, 36, 48, 60.

The airfoil of the blade used a pattern of two elements across
the thickness and eight elements along the chord. It was hy-
pothesized that the increase in the number of element layers
along the length would improve accuracy due to a highly complex, twisted, tapered airfoil surface. An increased number of

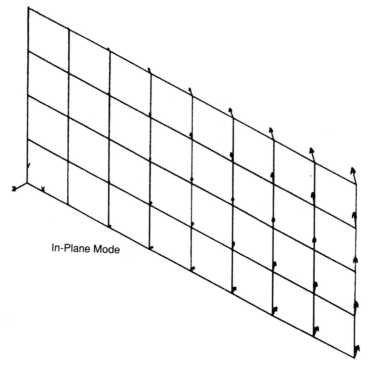

Figure 11.10   Flat plate example: Mode 4.

elements should increase the model's accuracy by providing a better representation of the blade's airfoil surfaces.

The results of frequency vs. number of dynamic degrees of freedom indicate, as shown in Table 11.6, that the first mode converged the fastest with the higher-order modes requiring more dynamic degrees of freedom to achieve convergence. All modes converged from higher frequencies down to lower frequencies. This was the same general trend exhibited by the cantilever plate model. A plot of frequency vs. number of dynamic degrees of freedom in Figure 11.13 shows the asymptotic convergence and illustrates that very little is gained in increasing the number of dynamic degrees of freedom beyond 36 for the first two modes. The third mode decreased by 0.4%

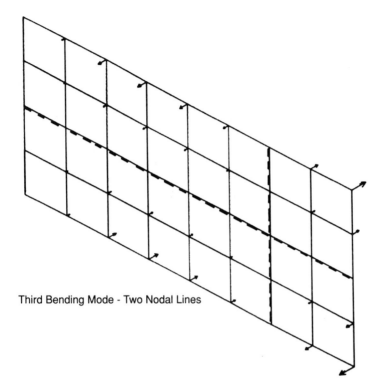

Third Bending Mode - Two Nodal Lines

Figure 11.11   Flat plate example: Mode 5.

when the dynamic degrees of freedom were increased from 36 to 60. The computer runtimes show an increase of 28% for the five-fold increase in dynamic degrees of freedom, from 12 to 60 (Tables 11.6 and 11.7).

The results of the variation in number of element layers from 6 to 30 showed similar results. All three modes converged downward from higher frequencies. The first mode had the fastest convergence, followed in order by the two higher modes. There was a larger spread in the computer runtimes of 41% between the 6-layer model (558 total elements) and the 30-layer model (942 total elements). This is to be expected because the element

Table 11.4 Comparison of Computer Runtimes for Plate and Solid Element Models—Cantilever Plate Example

32 Solid elements 90 Nodes 240 Total DOF			32 Plate elements 45 Nodes 200 Total DOF		

Householder eigenvalue solution

Element generation	24.37	sec. (0.762 sec/elem)	Element generation	29.54	sec. (0.798 sec/elem)
Guyan reduction	16.68	sec.	Guyan reduction	9.25	sec.
Eigenvalue solution 3 DDOF	0.221	sec.	Eigenvalue solution 3 DDOF	0.194	sec.
Misc.	14.61	sec.	Misc.	12.82	sec.
Total	55.88	sec.	Total	47.80	sec.

Full subspace iteration

Element generation	24.41	sec. (0.763 sec/elem)	Element generation	25.10	sec. (0.784 sec/elem)
Eigenvalue solution 4 modes	68.54	sec.	Eigenvalue solution 4 modes	82.25	sec.
Misc.	9.13	sec.	Misc.	1.07	sec.
Total	102.08	sec.	Total	108.42	sec.

Note: Computer runtimes are valid for a relative comparison between models and between cases.

formulation time, which is a large component of the total computer runtime, is directly proportional to the number of elements. A plot of frequency vs. number of element layers is shown in Figure 11.14 and data are given in Table 11.8.

Table 11.5 Comparison of Computer Runtimes for Various Combinations of DDOF—Cantilever Plate Example

32 Solid elements 90 Nodes 240 Total DOF		32 Plate elements 45 Nodes 200 Total DOF	
Guyan reduction	16.68 sec.	Guyan reduction	9.25 sec.
Eigenvalue solution (3 DDOF)	0.221 sec.	Eigenvalue solution (3 DDOF)	0.194 sec.
Guyan reduction	17.87 sec.	Guyan reduction	9.87 sec.
Eigenvalue solution (9 DDOF)	0.967 sec.	Eigenvalue solution (9 DDOF)	0.994 sec.
Guyan reduction	26.50 sec.	Guyan reduction	14.88 sec.
Eigenvalue solution (40 DDOF)	11.89 sec.	Eigenvalue solution (40 DDOF)	12.44 sec.

Note: Computer runtimes are valid for a relative comparison between models and between cases.

Mode shape plots of displacements of the dynamic degrees of freedom are shown in Figure 11.15 for the first three modes. The mode shapes are not that different from the simple plate. Because the airfoil resembles a tapered, twisted plate with much of its flexibility in the tip, the mode shapes are similar to the classical plate modes. The first mode is a simple plate-type bending mode. The second mode is orthogonal in the second most flexible direction, and the third mode is torsional.

The effects of stress stiffening on the steam turbine blade are shown in Table 11.9. Stress stiffening is sometimes called "piano wire effect"; it is the effect of a steady tensile stress on a slender, flexible structure that increases the structure's stiffness. The steam turbine blade is a relatively flexible structure that is subject to fairly high centrifugal stresses, up to 35 ksi, in the airfoil section. These stress-stiffening effects act to substantially raise the natural frequencies. This effect can be critical for designs such as this, or other rotating machinery

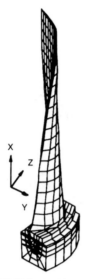

Y and Z DOF Used at Each Selected Node

Figure 11.12 Freestanding steam turbine blade.

Table 11.6 Modal Analysis Results of Solid Elements with Different Patterns of DDOF—Freestanding Steam Turbine Blade Model

Dynamic degree of freedom pattern	Frequencies, hz			CPU time, sec
	MODE1	MODE2	MODE3	
6 nodes/12 DDOF	179.7	412.1	725.1	4542.70
12 nodes/24 DDOF	179.6	409.1	714.8	4876.28
18 nodes/36 DDOF	179.0	407.3	709.7	5043.07
24 nodes/48 DDOF	179.0	407.2	708.0	5582.86
30 nodes/60 DDOF	179.0	407.1	706.6	5828.49

Note: Computer runtimes are valid for a relative comparison between models and between cases.

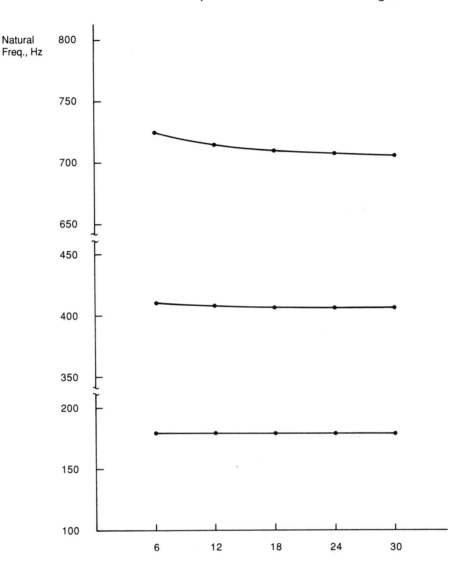

Figure 11.13 Natural frequency vs. number of dynamic degrees of freedom, 30 element layers.

applications, in which components must be carefully designed to prevent natural frequencies from coming into resonance with a potential forcing frequency.

Table 11.7 Computer Runtimes for Freestanding Steam Turbine Blade Example

744 Solid elements	
1227 Nodes	
3644 Total DOF	
36 Dynamic DOF	

Steady stress calculation	
Element formulation	640.91 sec. (0.861 ave.)
Displacement solution	1222.30 sec.
Element stresses	193.06 sec. (0.259 ave.)
Element forces	310.14 sec.
Misc.	455.97 sec.
Total	2822.38 sec.
Householder eigenvalue solution	
Element formulation	639.95 sec. (0.860 ave.)
Guyan reduction	3455.55 sec.
Eigenvalue solution	9.79 sec. (36 DDOF)
Misc.	44.09 sec.
Total	4149.38 sec.

Note: Computer runtimes are valid for a relative comparison between models and between cases.

11.4.4 Sheet Metal Box Modal Analysis Example

A sheet metal box was fabricated out of 10-gage (0.1345 in. thick) sheet steel and welded to a rigid base to use as a comparison for evaluating various finite-element models. The box is shown in Figure 11.16. A model was made using quadrilateral plate elements, with nine elements per panel (3 × 3) as shown in Figure

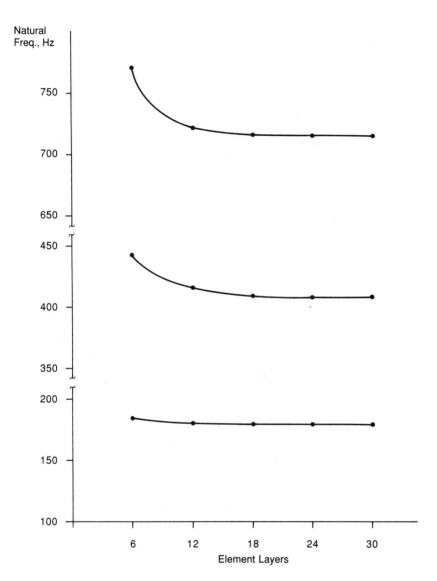

Figure 11.14 Natural frequency vs. number of vane elements along the length, 18 dynamic degrees of freedom.

Table 11.8 Modal Analysis Results of Solid Elements with Different Numbers of Element Layers—Freestanding Steam Turbine Blade Model

Number of element layers	Total number of elements	Frequencies, hz			CPU time, sec
		MODE1	MODE2	MODE3	
6	552	185.0	443.3	770.0	3452.77
12	648	179.9	415.6	721.6	3876.79
18	744	179.7	410.0	716.4	4149.38
24	840	179.7	409.5	715.6	4512.83
30	936	179.6	409.1	714.8	4876.28

Note: Computer runtimes are valid for a relative comparison between models and between cases.

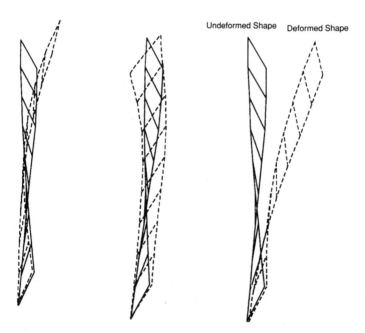

Undeformed Shape   Deformed Shape

Figure 11.15 Mode shapes of the freestanding steam turbine blade.

Table 11.9 Effects of Stress Stiffening on Natural Frequencies—
Freestanding Steam Turbine Blade Model, 18 Elements Layers,
36 Dynamic Degrees of Freedom

	Frequencies, hz		
	MODE 1	MODE 2	MODE 3
Without stress stiffening effects	179. 6	409. 1	714. 8
With stress stiffening effects-- rotation at 3600 rpm	222. 3	443. 0	749. 2

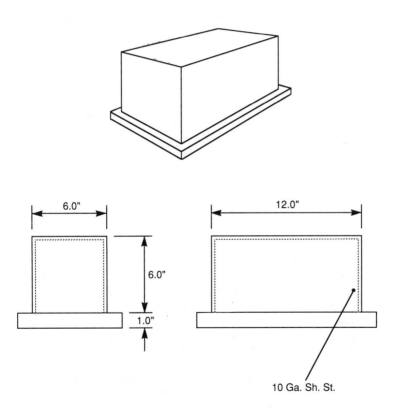

Figure 11.16 Sheet metal box used for modal analysis example.

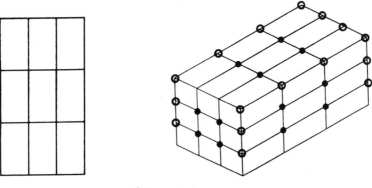

O Dynamic Degrees of Freedom Retained for First Model

● Dynamic Degrees of Freedom Retained for Second Model

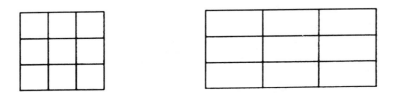

Figure 11.17 Finite-element model of sheet metal box.

11.17. Four separate finite-element calculations were performed and natural frequencies measured with an accelerometer and spectrum analyzer.

Two of the models were intended to investigate the significance of placement of dynamic degrees of freedom. The first model placed dynamic degrees of freedom along all of the edges and retained all six degrees of freedom per node. The second model placed dynamic degrees of freedom at the four interior nodes of each panel and retained the out-of-plane translation degree of freedom at each node. A Guyan reduction run was made to reduce the model to the dynamic degrees of freedom and the Householder method was used to solve the eigen problem. The results are given in Table 11.10. A comparison of numbers of dynamic degrees of freedom and computer runtimes is shown in Table 11.11.

Table 11.10 Modal Analysis Results of Sheet Metal Box Models

Model	Frequencies, hz				
	MODE1	MODE2	MODE3	MODE4	MODE5
Householder; edge DDOF	526.0	675.9	764.2	816.5	866.8
Householder; mid-panel DDOF	527.2	678.4	766.9	826.2	881.6
Modified Givens	516.0	685.2	797.3	861.2	886.9
Jacobbi - all DDOF	548.1	713.7	761.1	775.5	858.0
Test	515.0	680.0	710.0	835.0	

Table 11.11 Computer Runtimes for Sheet Metal Box

	First case	Second case
Model details:	45 Plate elements 40 Active nodes 480 Total active DOF 167 Dynamic DOF	45 Plate elements 40 Active nodes 220 Total active DOF 20 Dynamic DOF
Element formulation	33.203 sec. (0.738 ave.)	33.191 sec. (0.738 ave.)
Guyan reduction	23.382 sec.	25.130 sec.
Eigenvalue solution	110.194 sec.	1.609 sec.
Misc.	25.686 sec.	23.282 sec.
Total	192.465 sec.	83.212 sec.

Note: Computer runtimes are valid for a relative comparison between models and between cases.

The third calculation was performed with a different finite element program using the modified givens method and all degrees
of freedom. The fourth calculation was made with a third program using the Jacobbi method and all degrees of freedom. A
summary of all results is included in Table 11.9.

The results show that for the first two calculations, the placement of dynamic degrees of freedom at the center of the panels in
the out-of-plane direction gave almost the same results with about
half of the computer runtime. Plots of the mode shapes for the
first three modes are given in Figures 11.18–11.20. The mode
shape plots show that the modes are localized panel modes within each panel and that there are no overall bending modes of the
box that the edge degrees of freedom would pick up. These panel
modes have to be picked up by the rotational degrees of freedom

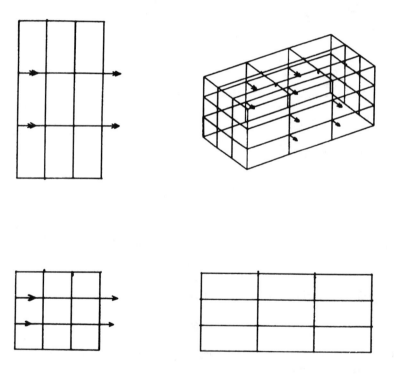

Figure 11.18 Sheet metal box: First mode.

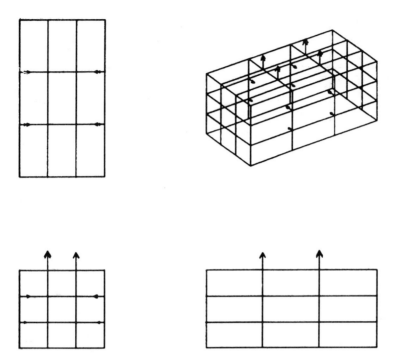

Figure 11.19 Sheet metal box: Second mode.

at the edge nodes. In addition, two of the first five modes were missed by the first calculation.

The comparison of the second, third, and fourth calculations and the experimental measurements showed that there is still a fair amount of difference between methods of calculation for the same model. This reinforces a basic concept, which is sometimes forgotten, that the finite element method is still a numerical approximation, even for a fairly simple problem such as this.

11.4.5 Combining Experimental Modal and Finite Element Modal Analysis

Natural frequencies and mode shapes can be measured experimentally using a modal analysis system consisting of a two-channel

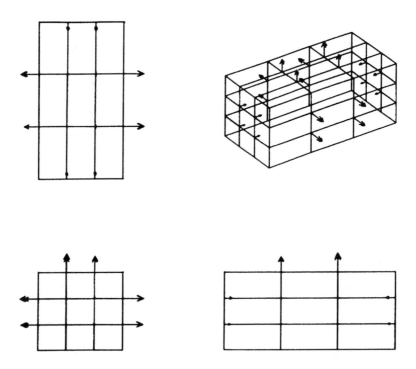

Figure 11.20 Sheet metal box: Third mode.

spectrum analyzer, a microprocessor/PC (often incorporated into the analyzer), and specialized software that can take measured transfer functions between a common reference point and a number of selected measurement points on a structure. A two-channel spectrum analyzer is used to acquire data simultaneously from both force input to the structure and response of the structure, generally measured with an accelerometer. A transfer function, the complex ratio of response to input force including the phase relationship between the signals, is calculated by the analyzer. These transfer functions are sent to a microprocessor that combines them with previously stored geometry data for the structure to display mode shapes for each of the predominant frequencies.

There is a close analogy between this procedure and the calculation of natural frequencies with a dynamic finite-element analysis model has been reduced to a selected set of dynamic degrees of freedom. The dynamic degrees of freedom are comparable to the experimental measurement points. The value of a comparison between experimental and analytical results is that the test results can be used to validate a finite-element analysis model before proceeding on to forced-response calculations. Experimental modal analysis, however, has the shortcoming of not allowing extensions to forced-harmonic or transient-response analysis, not being able to calculate dynamic stresses, and having very limited capabilities for predicting changes in dynamic response as a result of design changes to the structure.

11.5 TRANSIENT-RESPONSE ANALYSIS

Forced-response calculations are generally performed with the same reduced model as is used for modal analysis. It is a good rule of thumb to always perform a modal calculation prior to any forced-response calculations. Having the modal data, i.e., natural frequencies and mode shapes, is a virtual necessity for the interpretation of the forced-response results. With any forced-response calculation, data on damping and the applied force-time history need to be input along with the geometry and material property data previously input for modal analysis.

Transient-response analysis calculates the dynamic response of a structure due to a time-varying, but not periodic force. Typical transient forces are impact, step, or ramp functions. Direct time-integration schemes such as the Newmark beta, Houbolt, Wilson, or central difference methods are employed.

The same dynamic model employed for modal analysis is used for the transient-response calculation and the model is typically reduced to selected dynamic degrees of freedom prior to the actual transient-response calculation. Additional input for the transient-response calculation includes damping and specification of forcing that includes the time history of each of the dynamic degrees of freedom for which a force is applied. Each applied force may have a unique force-time profile and any time-phasing relationships must be accounted for in the force specification.

The results of a transient-response calculation for a reduced dynamic model are the displacement vs. time history for each of the dynamic degrees of freedom. The set of matrix equations has been given previously for transient response

$$\{-\omega^2 [M] + j\omega [C] + [K] \{X_0\} \sin(\omega t + \phi) = \{F(t)\}$$

The vector $\{X_0\}$ is the set of maximum displacements of the dynamic degrees of freedom.

The response of a structure to an impact and a step-forcing function is shown in Figures 11.21 and 11.22, respectively. These figures illustrate a simple beam reduced to a single dynamic degree of freedom. The step-forcing function is applied to the dynamic degree of freedom and the displacement response of the degree of freedom is calculated.

Once the transient response at the dynamic degrees of freedom has been calculated, the complete response of the structure may be obtained by back-substitution. The displacements of the dynamic degrees of freedom are used in conjunction with the back-substitution matrix generated during the dynamic condensation run, and the displacements of all degrees of freedom in the model are calculated for a specific time step. For the back-substitution step, the user must select the time step or steps for which the maximum displacements, and hence the maximum dynamic stresses, occur. Because the maximum displacements of all the dynamic degrees of freedom may not occur at the same time step, a certain amount of judgment must be used. After the back-substitution step has obtained the displacements of all the degrees of freedom in the model, the element stresses may be calculated. This set of stresses represents the instantaneous dynamic stresses corresponding to the selected time step. More than one time step may be selected for back-substitution; however, a full back-substitution run with stress calculation is time-consuming and roughly equivalent to a single static back-substitution and stress calculation.

11.6 FORCED HARMONIC-RESPONSE ANALYSIS

Forced harmonic-response calculations are generally performed with the same reduced model as used for modal or transient analysis. As in the case of transient analysis, it is recommended to always perform a modal calculation prior to any forced harmonic-response calculations. Having the modal data, i.e., natural frequencies and mode shapes, is a virtual necessity for the interpretation of the forced-response results. With any forced-response calculation, data on damping and the applied harmonic forcing need to be input along with the geometry and material property data previously input for modal analysis.

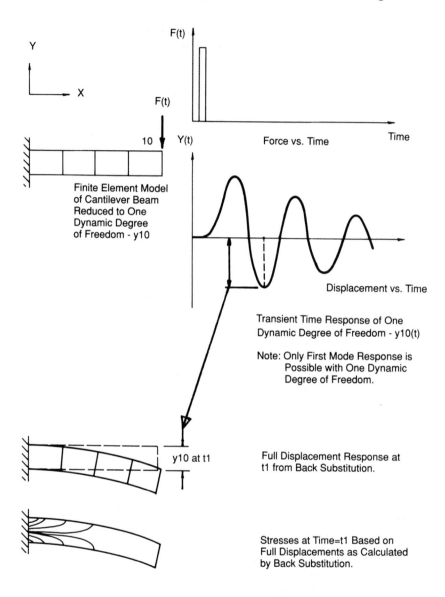

Finite Element Model
of Cantilever Beam
Reduced to One
Dynamic Degree
of Freedom - y10

Force vs. Time

Displacement vs. Time

Transient Time Response of One
Dynamic Degree of Freedom - y10(t)

Note: Only First Mode Response is
Possible with One Dynamic
Degree of Freedom.

y10 at t1

Full Displacement Response at
t1 from Back Substitution.

Stresses at Time=t1 Based on
Full Displacements as Calculated
by Back Substitution.

Figure 11.21 Transient response to an impact.

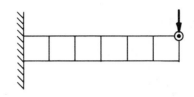

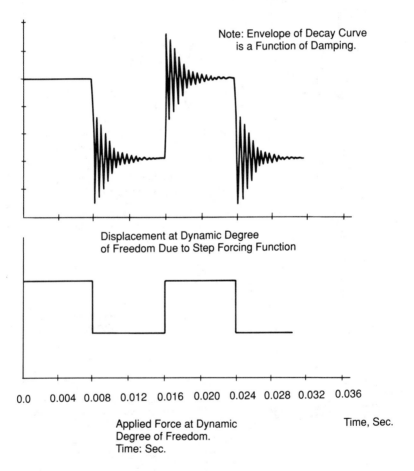

Note: Envelope of Decay Curve
is a Function of Damping.

Displacement at Dynamic Degree
of Freedom Due to Step Forcing Function

0.0   0.004 0.008 0.012   0.016   0.020   0.024 0.028 0.032   0.036

Applied Force at Dynamic   Time, Sec.
Degree of Freedom.
Time: Sec.

Figure 11.22 Step-forcing dynamic-response finite element model,
1 dynamic degree of freedom.

Forced harmonic-response (FRH) analysis is used to calculate the dynamic response of a structure to a periodic forcing function. FRH calculates the response of the structure to an applied sinusoidal forcing, one frequency at a time, so that for a non-sinusoidal, periodic forcing, it is necessary to first break down the forcing function via Fourier analysis into its harmonic components with appropriate phase relationships.

The same dynamic model as was used for modal analysis is employed for the FHR calculation, and the model is typically reduced to selected dynamic degrees of freedom prior to the actual FHR calculation. Additional input for the FHR calculation include damping and specification of forcing that includes frequency, force, amplitude, and relative phase angle. Force may be applied at any number of points. For a reduced dynamic model, forces may only be applied at dynamic degrees of freedom. All applied forces will be acting at the same frequency, although the relative phase angle and force amplitude may be unique.

The results of a FHR calculation for a reduced dynamic model are the maximum displacement and a relative phase angle at each of the dynamic degrees of freedom. The set of matrix equations has been given previously for FHR

$$\{-\omega^2 [M] + j\omega [C] + [K]\} \{X_0\} \sin(t + \phi) = \{F(t + \phi)\}$$

The vector $\{X_0\}$ is the set of maximum displacements of the dynamic degrees of freedom. This vector is often referred to as the "operational mode shape." This set of displacements will depend on the forcing frequency, i.e., it will resemble the free vibration mode shape of the next highest natural frequency. The operational mode shape will also be determined by the pattern of applied force.

The response of a structure to harmonic excitation can be shown, in Figure 11.2, by the dynamic amplification factor, X_0/F, which is the transfer function of the structure. The shape of the response curve at resonance is a function of the damping in that mode. With zero damping, the response at resonance would be infinite. The amount of damping determines the actual response amplitude as shown in the plot.

Once the harmonic displacement response at the dynamic degrees of freedom has been calculated, the complete response of the structure may be obtained by back-substitution. The displacements of the dynamic degrees of freedom are used in conjunction with the back-substitution matrix generated during the dynamic condensation run, and the displacements of all degrees of freedom in the model are calculated for a specific frequency and phase angle. Because the reduced FHR calculation gives the maximum amplitude for each dynamic degree of freedom regardless of phase angle, it is necessary, at the static back-substitution step, for the user to select a relative phase angle so that the dynamic degrees of freedom may be adjusted. After the back-substitution step has obtained the displacements of all the degrees of freedom in the model, the element stresses may be calculated. This set of stresses represents the instantaneous dynamic stresses corresponding to the selected phase angle.

11.6.1 Example: Forced Harmonic Response of a Simple Beam

Figure 11.23 shows a simply supported beam reduced to five dynamic degrees of freedom with a sinusoidal force applied to the midspan at a frequency of 11 Hz, which is just below the beam's first natural frequency of 12 Hz. Figure 11.24 shows the response of two of the dynamic degrees of freedom, numbers 1 and 2, along with the sinusoidal forcing function.

Figure 11.25 shows the response of the beam to a harmonic force applied at 49 Hz, just below the second mode frequency of 51 Hz. In order to better excite the beam near its second mode, equal forces of opposite direction (180° out of phase) are applied at dynamic degrees of freedom 2.

11.7 DAMPING

Damping is required for any forced-response calculations. In most programs, a simple representation of damping may be input, usually the damping ratio (ratio of actual damping to critical damping). In practice, damping is a combination of three mechanisms:

Figure 11.23 Simple beam: Harmonic-response example case.

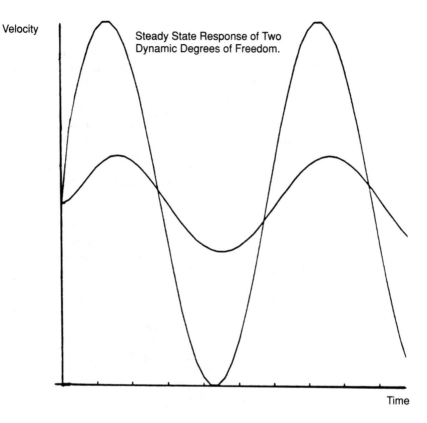

Velocity

Steady State Response of Two
Dynamic Degrees of Freedom.

Time

Figure 11.24 Simple beam: Response of 2 dynamic degrees of freedom.

Figure 11.25 Simple beam: Second mode response.

material or mysteretic damping due to internal material friction, coulomb or stick-slip damping due to dry friction between discrete parts of a structure, and aerodynamic damping due to viscous drag forces as the structure vibrates in air or a liquid.

Viscous damping is proportional to the vibration velocity, but the damping coefficient may not always be a constant. For example, material damping is known to vary according to dynamic stress, and for a structure vibrating near resonance with significant dynamic stress, this may account for a significant variation in material damping, causing the problem to become nonlinear. Damping also varies between modes. Higher modes tend to have a higher effective damping and die out faster from an impact.

A damping ratio may be obtained from either an experimental modal analysis of the structure to be modeled or a similar structure. In the absence of a complete modal test, a simple rap test of the structure with a storage oscilloscope can be used to obtain the first mode damping by capturing the decay trace from an impact and calculating the logarithmic decrement that is easily converted to a damping ratio number. A second alternative to the impact test is a sine sweep test in which the structure is excited over a continuous range of frequencies including one or more resonance frequencies. The shape of the response curve near resonance can be used to measure damping for each mode that is traversed. In particular, the bandwidth at the half-power points gives a good representation of damping. Details of these methods and conversions can be found in many vibration textbooks [2, 8, 9].

11.8 DYNAMIC SUBSTRUCTURING

Substructuring may be applied to dynamic models and the benefits are generally greater than for static analysis. With the use of dynamic condensation and reduced models, a substructuring-type procedure is already being applied to reduce the problem size prior to the actual dynamic calculations. When a repeated structure or a symmetric structure is to be analyzed, then dynamic substructuring is a necessary step.

The mathematics of dynamic substructuring are identical to the dynamic condensation procedures described in Chapter 2, and in Section 11.2.3. In addition to dynamic degrees of freedom, generally located in the interior of the model, boundary degrees of freedom must be retained in the dynamic condensation step in order to connect substructures. Often, after the substructures are generated and combined, a second dynamic condensation step may be used to eliminate the boundary retained degrees of freedom and leave only the interior dynamic degrees of freedom for the actual dynamic calculations.

An example of dynamic substructuring is the steam turbine blade discussed in Chapters 4 and 5. This turbine blade is intended to operate in groups of three joined together at the tip with a shroud. The static analysis model described in Chapter 5 showed a one-pitch segment of the shroud for the purpose of applying the appropriate centrifugal loading to the blade. In order to properly calculate the dynamic response of the turbine blade group, it is necessary to model the blade group. A dynamic substructure was formed as shown in Figure 11.26. The single blade substructure had 66 retained degrees of freedom, including 18 internal degrees of freedom (6 nodes, 3 DOF per node) and 48 boundary degrees of freedom (16 nodes, 3 DOF per node). When the substructure is replicated to form the three-blade group model, the 48 boundary DOF at the two interfaces would be condensed out and the dynamic calculations will be performed using the retained 18 DDOF in each of the airfoil sections for a total of 54 DDOF for the actual dynamic calculations. The assembled three-blade group model is shown schematically in Figure 11.27. The complete model, as shown in Figure 11.27 constructed from individual elements, never actually exists; however, this figure gives the effective model.

Runtimes for the various phases of the analysis of this model are given in Table 11.12. Because the three-blade group model is influenced by the centrifugal stresses, a steady-stress analysis precedes the substructure generation. The steady-stress calculation was discussed in detail in Chapter 5. Following the pattern shown in the previous examples, most of the computer time is taken up in the generation of the substructure that consists of element generation and dynamic condensation. The eigenvalue solution time is small by comparison. In the substructure use run, a second dynamic condensation pass is included to condense out the boundary degrees of freedom.

The use of substructuring has effectively reduced the problem from 11,349 degrees of freedom (3815 degrees of freedom per substructure times 3, less boundary degrees of freedom) down to 54 dynamic degrees of freedom; hence, the small amount of time actually required for the eigenvalue solution. The repetitious

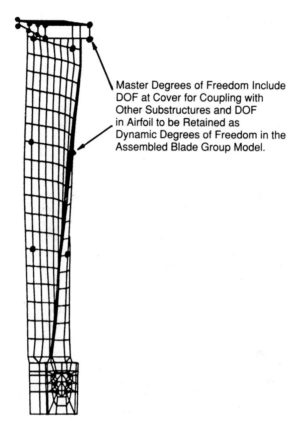

Master Degrees of Freedom Include
DOF at Cover for Coupling with
Other Substructures and DOF
in Airfoil to be Retained as
Dynamic Degrees of Freedom in the
Assembled Blade Group Model.

Figure 11.26 Turbine blade dynamic substructure.

element generation for the second and third blades in the model
has also been eliminated.

Mode shape plots of the reduced mode shapes, i.e., the displacements at the 54 dynamic degrees of freedom, are shown in
Figures 11.28–11.30 for the first three modes, respectively.

11.9 SUMMARY

Dynamic finite element analysis is a powerful tool for the anal-
sis of problems that might otherwise be impossible to obtain a

Master Degrees of Freedom Retained to Join
Substructures - To Be Removed in Second
Dynamic Condensation Pass on Assembled
Substructure Model

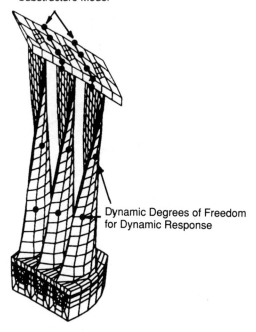

Dynamic Degrees of Freedom
for Dynamic Response

Figure 11.27 Turbine blade group: Assembly of substructures.

solution for. The modeling for dynamic analysis follows most of
the same basic rules as for static analysis, and the same model
will generally prove satisfactory for both. Several items can be
listed to summarize this chapter.

    Any dynamic analysis should include modal analysis as the
    first step. Identification of natural frequencies and mode
    shapes is fundamental in any dynamic analysis. Knowing
    where the natural frequencies are and what the mode shapes
    look like is important information before a transient or forced-
    harmonic calculation.
    The user should have a qualitative feel for the dynamic be-
    havior of the structure in the same way he anticipates the
    static deflection and stress response.

Table 11.12 Computer Runtimes for Steam Turbine Blade Group
Dynamic Substructuring Example

Substructure:	Global model:
774 Solid elements	3 Repeated substructures
1272 Nodes	66 Master DOF / Substruct.
3815 Total DOF	54 Retained dynamic DOF
66 Master DOF	

Steady stress calculation	
Element formulation	665.89 (0.860 ave.)
Displacement solution	930.55
Element stresses	200.97 (0.260 ave.)
Element forces	14.37
Misc.	445.59
Total	2257.37 sec.
Substructure generation	
Element formulation	668.16 (0.863 ave.)
Guyan reduction	4628.58
Write files	240.37
Misc.	274.49
Total	5811.60 sec.
Householder eigenvalue solution	
Read in data	66.88
Guyan reduction	17.90
Eigenvalue solution	21.87
Misc.	91.99
Total	198.64 sec.
Analysis total	8267.61 sec.

Note: Computer runtimes are valid for a relative comparison between models and between cases.

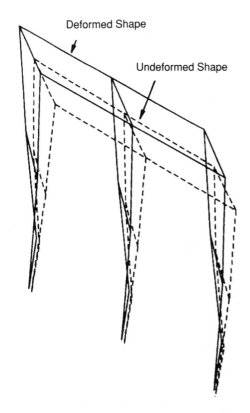

Deformed Shape

Undeformed Shape

Figure 11.28 Turbine blade group: First mode.

Dynamic geometry modeling is often very similar to geometry
modeling for static analysis. Symmetry cannot be used in
dynamics the way it can in static models. Dynamic substructuring can be used to help overcome this problem.
The single greatest difference between static and dynamic
modeling is that dynamic modeling requires a "second" model
to be defined, which is the subset of degrees of freedom to
be used in the actual calculation. The results of a dynamic
calculation will be influenced more by the placement and number of dynamic degrees of freedom than will be the original
global model.

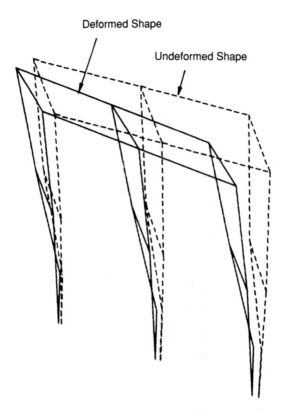

Figure 11.29 Turbine blade group: Second mode.

Dynamic finite element analysis is a powerful tool that requires knowledge in the fundamentals of vibrations and knowledge of finite element techniques. Of the two, the knowledge of vibration fundamentals is probably the most important and requires the most experience. In addition, anyone having a familiarity with linear static finite element analysis should not hesitate to try dynamic analysis.

Deformed Shape

Undeformed Shape

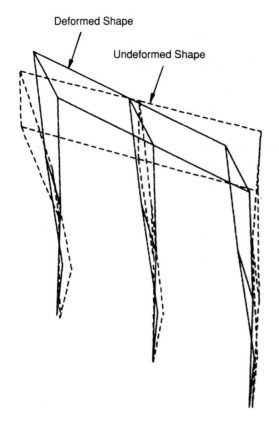

Figure 11.30 Turbine blade group: Third mode.

REFERENCES

1. Roark, R. J. and W. C. Young, *Formulas for Stress and Strain*, 5th ed., McGraw-Hill, New York, 1975.
2. Timoshenko, S. P., D. H. Young, and W. Weaver, *Vibration Problems in Engineering*, 4th ed., Wiley, New York, 1974.
3. Harris, C. M., *The Shock and Vibration Handbook*, 3rd ed., McGraw-Hill, New York, 1988.
4. Bittner, J. L., "ANSYS Revision 4.3 Tutorial–Newmark Integration Solution Technique," Swanson Analysis Systems, Inc., Houston, PA, DN-T003, March 1987.

5. Guyan, R. J., "Reduction of Stiffness and Mass Matrices," *AIAA Journal*, Vol. 3, No. 2, Feb. 1965.
6. Zienkiewicz, O. C., *The Finite Element Method in Engineering Science*, 2nd ed., McGraw-Hill, London, 1971.
7. Steele, J. M., "Finite Element Modal Analysis of Steam Turbine Blades," Proc. 4th International Modal Analysis Conf., Los Angeles, Calif., Feb. 1986.
8. Steidel, R. F., *An Introduction to Mechanical Vibrations*, Wiley, New York, 1971.
9. Baumeister, T., ed., *Standard Handbook for Mechanical Engineers*, 7th ed., McGraw-Hill, New York, 1967.

12

THERMAL ANALYSIS WITH FINITE ELEMENT ANALYSIS

Thermal analysis is one of the most straightforward applications of finite element analysis and is in many respects a more simple application than the common static stress/displacement analysis. Complexity and difficulty in thermal analysis arise in two areas: (1) quantifying boundary conditions, especially convection boundary conditions, and (2) the computer time required for nonlinear solutions of radiation and convection problems.

Although finite element analysis heat-transfer calculations are useful by themselves, most interest is in the ability to calculate thermally induced displacements and stresses. Thermal displacements and stresses can be caused by several situations: (1) differential expansion of dissimilar materials in the same component, (2) restricted expansion due to a rigid constraint, and (3) transient stresses due to nonuniform temperature gradients.

12.1 PRIMER ON HEAT TRANSFER

Heat transfer can be divided into three basic mechanisms: conduction, convection, and radiation. Like all finite element problems, the units of thermal analysis are entirely arbitrary so long as they are consistent. In this chapter, however, the English units are given for clarification of some of the more complex coefficients.

12.1.1 Conduction

Conduction refers to the transfer of heat through a solid or unmoving fluid and is governed by a linear equation

$$Q = \frac{k\,A}{L}\,T$$

where

Q = heat flux rate, BTU/hr
k = material conductivity, BTU/in.-hr-°F
A = cross-sectional area of the conducting path, in.^{2}
L = length of the conducting path, in.
T = temperature difference across the conducting path, °F

It can be seen that the rate of heat conduction is proportional to the cross-sectional area and inversely proportional to the length of the path of conduction. The rate of conduction is linearly proportional to the temperature difference that lends itself to a simple solution that can be readily adapted to a finite element formulation.

12.1.2 Convection

Convection is the transfer of heat through a fluid where the fluid is moving. Convection can either be forced, when one fluid is pumped over a surface such as in a heat exchanger, or free, when the fluid circulates due to convection currents alone. An expression for free convection from a hot surface is

$$Q = h(T_f)\ A\ (T_s - T_{inf})$$

where

Q = heat flux rate BTU/hr
T_s = surface temperature

T_{inf} = freestream fluid temperature, assumed to remain constant

T_f = film temperature, generally the average of the surface
and freestream temperatures

$h(t_f)$ = convection film coefficient that is a function of the film
temperature

A = surface area

It can be seen that even if $h(t_f)$ is a linear function, the problem is still indeterminate because the convection film coefficient
$h(t_f)$ is a function of the surface temperature. In finite element
models, convection is used as a boundary condition to conduction
problems so that the surface temperature is an unknown. The
surface temperature must then be solved for so that both the
conduction and convection conditions are satisfied.

12.1.3 Radiation

Radiation is the transfer of heat either through a vacuum or air.
Radiation and convection may take place in simultaneously and
independently between two surfaces. The general form of an
equation for radiation heat transfer between two parallel surfaces is

$$Q = \sigma A(T_h^4 - T_c^4)$$

where

Q = heat flux rate, BTU/hr

σ = Boltzmann's constant 0.1714×10^{-8} BTU/h-ft^2-deg^4

A = surface area, if we assume that the areas of both surfaces are equal and all the energy leaving one surface
is absorbed by the other

T_h = temperature of the hotter surface in °R (+460°F)

T_c = temperature of the colder surface in °R

It should be noted that the heat flow rate is dependent on the
temperatures to the fourth power and not the temperature difference to the fourth power. Therefore, radiation can be predominant at high temperatures, i.e., above 1000°F, but may be
negligible at room temperature. Radiation also requires a nonlinear finite element formulation because of the fourth-power temperature terms.

12.2 THERMAL FINITE ELEMENT ANALYSIS

Thermal finite element analysis can be generally broken down into three main areas: (1) steady-state heat transfer, (2) transient heat transfer, and (3) thermally induced displacements
and stresses.

The governing matrix equations for steady-state conduction
are fairly simple. The basic set of matrix equations governing
simple conduction can be taken from the Laplace relationship.
The matrix set of equations is

$$[C] \{T\} = \{Q\}$$

where

[C] = global conductivity matrix
{T} = nodal temperature vector, one "degree of freedom" per
    node, temperature
{Q} = nodal heat flux vector

Although heat flux flow is across the entire element boundary, it
is resolved into nodal loadings in the same way as distributed
pressure is resolved into nodal forces.

When convection boundary conditions are specified, the set of
matrix equations is modified

$$[C] \{T\} = \{Q\} + \{h(T)\} \{T\}$$

where {h(T)} is the set of convection coefficients for the appropriate boundary nodes. The default condition for external boundaries is perfect insulation.

For transient heat conduction, the governing time-dependent
equation is

$$K \frac{d^2T}{dx^2} + \frac{\rho c}{k} \frac{\partial T}{\partial t} = 0$$

where

ρ = material mass density, lbm/in.^{3}
c = material specific heat, BTU/lbm-°F
k = material conductivity, BTU/in.-h-°F

Various time-marching procedures exist for implementing the time-
dependent finite element solution.

Nodal temperatures are included in the static displacement and
stress solution in the form of an equivalent set of nodal loadings

$$[k] \{d\} = \{F\} + \{F_T(T)\}$$

where

$$\{F^T(T)\} = \int [B][D][\alpha] \, dvol \, \{T\}$$

and

$[\alpha]$ = identity matrix of the thermal coefficient of expansion

$[B][D]$ = matrix product that relates nodal displacements to nodal forces

The thermal force vector $\{F_T(T)\}$ can be added to an applied force vector $\{F\}$ to calculate a combined loading or used separately to calculate thermal displacements and stresses alone. In performing analyses in which thermal effects are to be combined with external forces, it is advisable to perform three calculations: thermal effects alone, applied external forces alone, and combined thermal and external forces. This will give a good indication of the relative effects as well as the combined result.

12.3 THERMAL MODELING CONSIDERATION

In many cases, it is desired to use the results of a thermal calculation as input to a structural run. In these cases, one model may be developed with the structural run in mind and sufficient detail included for stress concentration factors, etc. In thermal models, less detail is required and a coarser model may be sufficient. Details such as notches and fillets will probably not affect the temperature distribution as much as they will the stress profiles. In the case of notches or corners, any thermal affect may be on the convection boundary conditions that will be difficult to quantify in any event. Unless details of the convection conditions, i.e., fluid temperature and velocities are known, the model will probably assume the same conditions in the corner or notch as on an adjacent outside surface.

The type of thermal analysis will dictate whether or not any modification to the structural model is necessary. For linear steady-state cases, the thermal model will run in less time than the structural model because of the single degree of freedom per node, so that using the same model for both analyses is the easiest way to go. For highly nonlinear problems such as radiation, nonlinear convection or a thermal transient calculation of a coarse thermal model may be justified, especially if the structural model contains small details needed for stress concentration effects. In these cases, a refined mesh modeling approach to the structural model may be in order to still allow one basic geometry model to be used for both the thermal and structural models.

There is always the problem of transferring temperature data from the thermal run to the structural run if the same geometry model is not used. In cases where the same model is used, the transfer is generally handled automatically within the finite element program. When the same model is not used, the user is faced with interpolating by hand or writing a separate interpolation program to obtain temperature data for the structural model.

For conduction problems, a large number of interior elements are not required. Because conduction is linear, the temperature gradient between two points will be constant and a minimal number of elements are required.

The most difficult part of thermal modeling is not the generation of the model or specification of boundary conditions, but knowing what boundary conditions to specify. For radiation problems, the unknowns are the material emmisivity and view factors. View factors are a geometry problem and can be estimated from tables [1] or calculated. Some commercial finite element codes have a view factor capability built in or they have a companion program. Material emmisivity is critical to radiation analysis and is a function of the material, surface finish, and temperature. Heat-transfer texts such as Holman [1] give tables of emmisivities for various materials.

Convection, both natural and forced, is a complex phenomenon involving details of the fluid flow over a solid surface. The fluid's density, conductivity, coefficient of thermal expansion, and viscosity all are involved. The surface finish of the solid surface also may be involved as it affects the boundary layer. The effective rate of forced convection is strongly dependent on the fluid velocity and where the fluid is stalled; for example, in corners or notches, the convection rate may be much lower than along a free surface. Most finite element programs allow only a simple linear or first-order nonlinear input of a net convection coefficient or range of coefficients. Therefore, some simplifying assumptions must be made ahead of time to determine the convection coefficients to be used. If the resulting surface temperatures are significantly different from those assumed in the premodeling convection calculations, then some adjustment and manual iteration may be necessary.

12.4 EXAMPLE: PRESSURE VESSEL

The pressure vessel example problem discussed in Chapters 4 and 5, Sections 4.2.3 and 5.4.3, is modified to include the effects of a nonuniform temperature distribution. Figure 12.1 shows the axisymmetric pressure vessel. The pressure vessel is made of two different materials with different conductivities and coefficients of thermal expansion. Material 1 has a conductivity of 15.0 BTU/in.-°F and a coefficient of thermal expansion of 9.0×10^{-6} in./in./°F. Material 2 has a conductivity of 25.0 BTU/in.-°F and a coefficient of thermal expansion of 7.0×10^{-6} in./in./°F

The contents of the pressure vessel are assumed to be at a uniform temperature of 500°F. Natural convection between the vessel contents and the inside of the vessel wall is assumed with a constant convection coefficient of 50.0 BTU/in.^{2}-°F-h. The outside of the pressure vessel is perfectly insulated on the side, but allows for heat loss (free convection) out of the end of the pressure vessel with a constant convection coefficient of 30 BTU/in.^{2}-°F-h and an ambient outside temperature of 100°F. Figure 12.1 shows the pressure vessel with the boundary conditions indicated. Figure 12.2 shows the fine-mesh finite-element model with 150 elements and 204 nodes.

The boundary conditions were specified as follows:

Perfect insulation Default, no boundary conditions specified.

Inside of pressure vessel Convection boundary conditions applied to all interior nodes with a constant ambient temperature of 500°F and a constant convection coefficient of 50 BTU/in.^{2}-°F-h.

Outside right-hand end of vessel Convection boundary conditions applied with a constant ambient temperature of 100°F and a constant convection coefficient of 30 BTU/in.^{2}-°F-h.

Two separate calculations were run using the fine gridwork in Chapter 5, Section 5.4.3. The first calculation was a steady-state thermal run to calculate the internal node temperature distributions, and the second was a structural run to calculate displacements and stresses. The thermal run was a steady-state calculation using the boundary conditions specified above. It was assumed that the temperatures of the contents of the pressure vessel or the ambient air surrounding the vessel were constant and not affected by the heat-transfer process. In the structural run, both pressure and thermal effects were included.

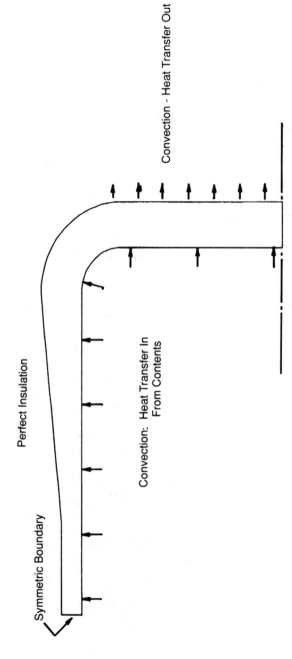

Figure 12.1 Axisymmetric pressure vessel with boundary conditions.

Figure 12.2 Fine-mesh pressure vessel model.

Table 12.1 gives the computer runtimes for the two calculations. The thermal run is much faster than the structural run
simply because in the thermal run, there is only one degree of
freedom per node. In the thermal run, element and nodal heat
fluxes were calculated, which was not necessary if only nodal
temperatures were required. There is almost a 2:1 difference
in the runtimes that is about the ratio of degrees of freedom.
This thermal/structural run is very similar to the structural
run with the fine model discussed in Chapter 5 and given in
Tables 5.5 and 5.6. The difference is that in the thermal/struc-
tural run, the nodal temperatures from the thermal calculation
were pulled in and an equivalent loading due to the thermal expansion was added to the pressure loading to give the composite
displacements and stresses.

The stress results of the thermal/structural run are given in
Table 12.2, along with the results of the Chapter 5 pressure-
only stresses, for comparison. As is evident by this comparison, the thermal effects contribute to a much higher stress than
the pressure-loading alone. At the point of maximum stress, the
major contributor to the stress is the steep temperature gradient in the pressure vessel at the point where the insulation is
removed. The higher stress is due also to the difference in the
coefficients of thermal expansion of the two materials and the

Table 12.1 Details and Computer Runtimes for Pressure Vessel Thermal and Structural Models

---

Thermal (fine model)
  150 Elements (two-dimensional, axisymmetric)
  204 Nodes   204 active DOF
  Times (sec)

Element formulation	6.176 (0.041 sec/elem.)
Wavefront solution	2.058
Element heat flux	1.800 (0.012 sec/elem.)
Nodal heat flux	4.503 (0.030 sec/elem.)
Misc.	27.672
Total	42.209 sec

Structural (fine model)
  150 Elements (two-dimensional, axisymmetric)
  204 Nodes   400 active DOF
  Times (sec)

Element formulation	47.267 (0.315 sec/elem.)
Wavefront solution	7.042
Stress solution	17.627 (0.118 sec/elem.)
Element forces	1.130
Misc.	26.925
Total	99.991 sec

---

fact that they are specified as being rigidly joined at their interface.

A plot of the temperature profile is shown in Figure 12.3. The thermal flux vectors are illustrated in Figure 12.4. This plot shows that all the heat flows out of the pressure vessel at the right-hand end where there is no insulation and convection is specified. As previously mentioned, perfect insulation is the default in any thermal model, and it is specified simply by not specifying any boundary condition along the selected surfaces.

A plot of maximum principal stress for the combined thermal/structural calculation is shown in Figure 12.5 with the maximum principal stress plot from the presure-loading-only run included for comparison. It must be noted that different scales

Table 12.2 Stress Results of Thermal Pressure Vessel Models

	Maximum centroidal stress				Maximum nodal stress			
	σ_r	σ_a	σ_t	σ_1	σ_r	σ_a	σ_t	σ_1
Pressure loading only (fine model)	594.18	565.04	235.47	1096.6	943.13	991.78	436.95	1836.16
Pressure and thermal loading (fine model)	20,213.0	30,492.0	-301.00	30,492.0	36,119.0	42,528.0	-1324.7	42,528.0

σ_r = radial stress
σ_a = axial stress
σ_t = tangential (hoop) stress
σ_1 = maximum principle stress

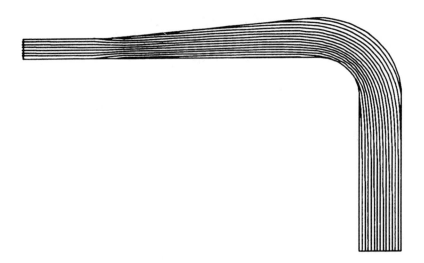

Figure 12.3 Steady-state temperature profile plot.

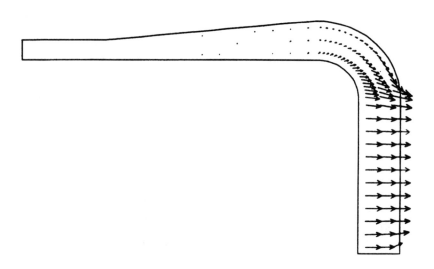

Figure 12.4 Thermal flux vector plot.

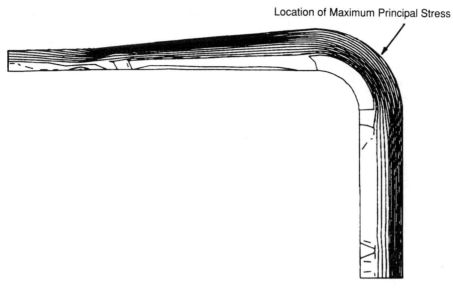

Figure 12.5 Stress contour plot—thermal and pressure loading.

are used for the two plots. The main purpose of the plots is to show the differences in stress distribution.

12.5 SUMMARY

Thermal finite element analysis is relatively straightforward by itself. The two complicating matters are knowing and specifying the appropriate boundary conditions and the fact that many calculations are nonlinear, requiring substantial computer runtime. For linear, steady-state calculations, the computer runtime is less than for a comparable structural model because the thermal model has only one degree of freedom, temperature, per node as compared to two or three degrees of freedom for a structural model.

When thermal effects are present, they may drastically affect the stress and displacement results and should be accounted for. As a simple example, if a piece of steel is rigidly constrained and heated by a moderate 100°F, the resulting expansion and strain will be about 7×10^{-4} in./in. (7×10^{-6} in./in./°F $\times$ 100°F), resulting in a stress of 21,000 psi. It is easy to imagine other real-life combinations of constraints and temperature differentials that could also lead to stresses which are in excess of those generated by applied loading.

Many commercial finite element programs have both thermal and structural capability and allow for efficient transfer of temperature data from one run to another as well as convenient restart capabilities.

REFERENCE

1. Holman, J. P., *Heat Transfer*, 3rd ed., McGraw-Hill, New York, 1972.

13

CALIBRATING THE ACCURACY OF FINITE ELEMENT MODELS

Probably the most difficult part of any finite element analysis is making an estimate of the absolute accuracy of the results. This part of an analysis requires skill and experience both in the finite-element method and with understanding the actual behavior of the structure under analysis. Although a set of "industry standard" benchmark problems is being agreed on [1], the generally available data for comparison of the accuracy of different element types are for fairly simple models with controlled element shapes and clearly defined boundary conditions. Problems offered in the verification manuals of various finite element codes have the same drawback of showing simple shapes under ideal conditions.

Calibrating a real problem with semidistorted elements and partially unknown boundary conditions is a more difficult job. Finite element models are generally constructed using nominal dimensions and physical properties. However, all of the physical parameters have a certain allowable tolerance on them. Quantifying parameters such as the stiffness between mating parts,

for example, is compounded by variations in assembly procedures
and effects of operation.

Establishing acceptable limits of absolute accuracy is subjec-
tive, at best. In one industry, acceptable accuracy might be
±20%, a number that would be completely unacceptable in another
industry.

13.1 SOURCES OF UNCERTAINTY

Discrepancy between finite element results and actual behavior
can be divided into two categories: (1) finite element uncertainty, i.e., differences between nominal results and finite element calculations, and (2) manufacturing variability, variations
in actual results due to manufacturing tolerances, assembly variations, and operating conditions. These categories of uncertainty
can be subdivided into three areas: (1) model geometry, (2)
boundary constraint conditions, and (3) applied forces.

13.1.1 Model Geometry: Finite Element Uncertainty

Finite element uncertainty may be due to insufficient definition
of structural geometry resulting from too few elements. Also,
when there are high-stress gradients, the peak stress may not
be accurately determined unless there are a sufficient number of
elements to define the gradient. This problem usually shows up
as large differences in stress between adjacent elements. Element distortion and numerical problems may result from attempting to use too few elements to conform to a complex geometry.
For example, in matching a curved surface with straight-sided
elements, it is obvious that the model will underpredict the actual shape on an outside radius and overpredict the shape on an
inside radius. The model will approach the true shape as the
number of elements increases. This is illustrated in Figure 13.1
showing a circular shaft in bending. In this simple example,
three cases are run with 6, 10, and 20 elements in the cross
section. The results in Figure 13.1 show that the six-element
case is understiff due to the fact that the total cross-sectional
area is less than the true area.

Another problem associated with the element density of a
model is element distortion. In models of the generator cross
section in Figure 13.2, it may not be possible to use perfectly
rectangular elements with 90° vertices. In such cases, the

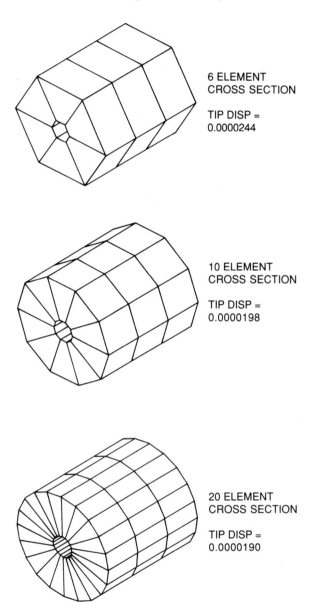

6 ELEMENT
CROSS SECTION

TIP DISP =
0.0000244

10 ELEMENT
CROSS SECTION

TIP DISP =
0.0000198

20 ELEMENT
CROSS SECTION

TIP DISP =
0.0000190

Figure 13.1 Circular shaft: Three models.

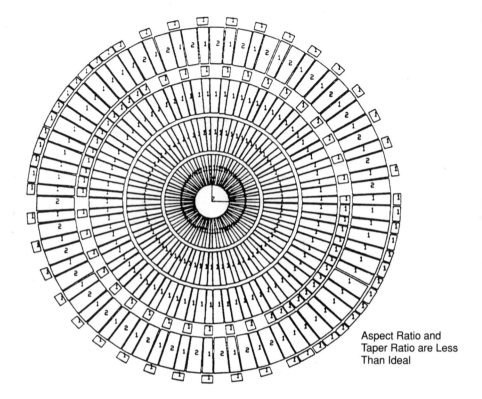

Aspect Ratio and
Taper Ratio are Less
Than Ideal

Figure 13.2 Generator cross section: Semidistorted elements.

degree of distortion is inversely proportional to the element density.

In dynamics problems, the number of master degrees of freedom affects the accuracy of the calculated natural frequencies and the dynamic response. As a rule of thumb, setting the total master degrees of freedom equal to twice the highest mode of interest will tend to eliminate this source of discrepancy. This topic is discussed in detail in Chapter 11.

13.1.2 Manufacturing Variations

A second source of uncertainty is in the deviation of actual dimensions from the nominal dimensions. In some cases, the

nominal dimensions used to form the finite-element model may
not represent the actual component. More important are the
fit between parts, tightness of bolts, and quality of welds,
etc. This may be of particular importance in failure investi-
gations where a stress-related failure may be caused by deviation, either within or beyond accepted limits, from nominal
specifications.

13.1.3 Boundary Conditions

Boundary constraints require assumptions as to the basic behavior of the structure. One determination in the formation of
any finite element model is how much of the structure and supports to include in the model. For example, in the case of the
bracket shown in Figure 13.3, a decision has to be made as to
where to consider the structure to be rigidly constrained. If
the frame at point B is significantly more rigid (a factor of 10
would be preferred) than the bracket itself, then specification
of rigid constraints at the interface would be sufficient, thereby minimizing the size of the model. However, if the frame at
B were flexible, i.e., a sheet metal fabrication rather than a
casting, then part of the frame would have to be included in
the model to a point where it is safe to assume that displacements in the frame no longer affect the stresses at the hole A.
    A second source of discrepancy between finite element pre-
dictions and actual behavior is assumed internal conditions.
For example, in machinery there may be bolted connections that
are assumed to be rigid when, in fact, they may have become
loose in operation. In thermal problems, internal heat conduction may be complicated by the interface between discrete parts
where there is rarely a perfect interface.
    An example of the combined effects of the above can be given
by the comparison of modal analysis results of a steam turbine
blade row [2]. In this case, 88 blades were grouped into 22
groups of 4 by joining them at their tips with a cover (Figure
13.4). A model was developed for a blade using solid elements
(Figure 13.5). A substructure was formed with the single blade
model and replicated to form the blade group model. Natural
frequency and mode shape calculations were performed. Experimental modal analysis tests were performed on the blade
groups. Measured natural frequencies were obtained for all
22 blade groups. There was a significant amount of scatter
within the measured data. Figure 13.6 shows the scatter in

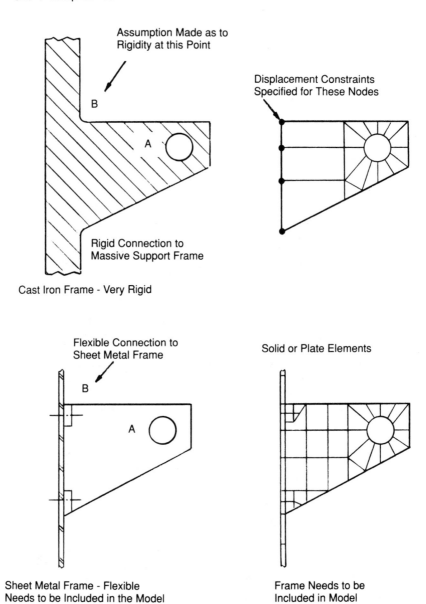

Assumption Made as to
Rigidity at this Point

B

A

Rigid Connection to
Massive Support Frame

Cast Iron Frame - Very Rigid

Displacement Constraints
Specified for These Nodes

Flexible Connection to
Sheet Metal Frame

B

A

Solid or Plate Elements

Sheet Metal Frame - Flexible
Needs to be Included in the Model

Frame Needs to be
Included in Model

Figure 13.3 Bracket attached to two supporting structures.

Figure 13.4 Steam turbine blade used for comparison of experimental modal results with finite-element calculations.

natural frequencies for the first three modes and the calculated natural frequencies. This example shows that if we consider the amount of scatter in the actual results, the comparison with the calculated results is reasonable and the model can be considered "nominally acceptable." However, the amount of scatter in the results also illustrates the point that a single natural frequency value for each mode would be inaccurate and

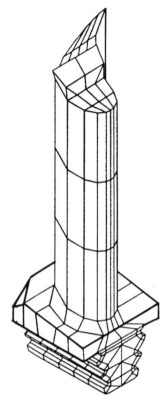

Figure 13.5 Three-dimensional finite element model of steam turbine blade.

13.1.4 Applied Forces

Selecting and quantifying the forces to be applied to a model can
be the most difficult part of an analysis. Applied forces can include static point loads, static distributed loads, transient impact loads, harmonic vibratory loads, and thermal forces due to
differential expansion. Forces must be applied with the correct
magnitude, distribution, points of application, and in some cases
frequency and phasing.

NUMBER OF
OCCURRENCES

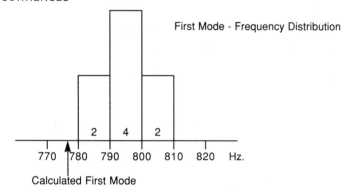

First Mode - Frequency Distribution

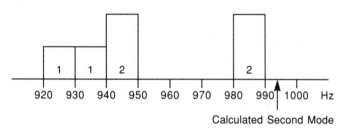

Second Mode - Frequency Distribution

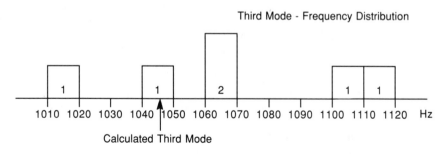

Third Mode - Frequency Distribution

Figure 13.6 Distribution of Modal Test Results: First three modes.

13.2 METHODS OF CALIBRATION

The calibration of finite element results may be broken down into deterministic and probabilistic methods:

Deterministic Comparison with test results from an actual structure.

Probabilistic Mathematically bounding the problem with sensitivity analysis.

13.2.1 Deterministic Approach: Comparison of Test Results

The most basic way of checking the accuracy of an analysis is by comparison with test data from the actual structure. Measurement techniques include the use of dial indicators and strain gages for deflections and stresses, accelerometers for the measurement of dynamic response, thermocouples for steady-state and transient temperature distributions. For determination of vibration mode shapes, experimental modal analysis equipment may be used. Simple techniques for obtaining an overall stress or temperature distribution include the use of brittle coatings and temperature-indicating coatings, respectively.

Test results can help to establish loading conditions as well as calibration of the finite element analysis. For example, in the case of rotating machinery, the assumption may be made that the only applied forces are centrifugal forces. The finite element analysis will then give stresses that are linearly proportional to rotational speed squared, and zero at zero speed. If test results show stresses that are not perfectly linear with speed squared, then this may indicate other external forces such as drag due to seals acting on the machine. If the measured stresses do not extrapolate to zero at zero speed, then this is an indication of instrumentation offset or thermal expansion effects that need to be removed from the test results before comparison with finite element results. When the influence of additional forces is found, it may be removed from the test results by regression analysis for the purpose of verifying the finite element model and then added as input to the finite-element model for an updated calculation.

Test techniques have the disadvantage of displaying data at only one point in the case of strain gages, pressure transducers, thermocouples, or accelerometers. This can also present problems in cases of high stress or temperature gradients. In setting up a test for comparison with finite element calculations, data at several points in the structure should be collected over a wide range of operating conditions. This allows for quantification of basic trends and identifies any malfunctioning transducers. In the event that poor correlation is found between test data and finite element results in the first comparison, the calculated stress and temperature gradients should be checked. In many cases, this discrepancy may be attributed to not obtaining test (strain gage or thermocouple) and finite element results at precisely the same location and in the same direction.

13. 2. 2 Probabilistic Approach: Mathematically Bounding the Problem

Both the finite element uncertainty and the uncertainty due to manufacturing variance may be partially clarified without the availability of actual test data by bounding the input data and performing a series of parametric studies. In order to completely quantify the uncertainty of any finite element result, a true probabilistic approach would be required. This would include establishing the statistical distribution of all input parameters such as geometry, applied forcing, rigidity of boundary conditions, etc. These distributions for each of the input parameters should be tested with the finite element model and the results combined using an appropriate combination law. An example of the final result can be shown in Figure 13. 7 for the combination of calculated steady stress and allowable yield strength. A complete discussion of probablistic methods is beyond the scope of the present topic, but it is given in a number of sources such as (3) and (4).

An alternative approach is to bound each of the input parameters with reasonable limits and proceed to thereby test the sensitivity and bounds of the results. In many cases, it may be difficult to establish the statistical or probabilistic distribution of an input parameter such as a distributed force; however, it may be possible to establish reasonable upper- and lowerbound numbers.

13. 2. 3 Bounding Finite Element Uncertainty

This upper/lowerbound approach may be applied to the finite element uncertainty as well as the manufacturing uncertainty. In order to test and bound the effects of model geometry and mesh density, the rate of convergence of a particular model may be established by running several cases with different mesh

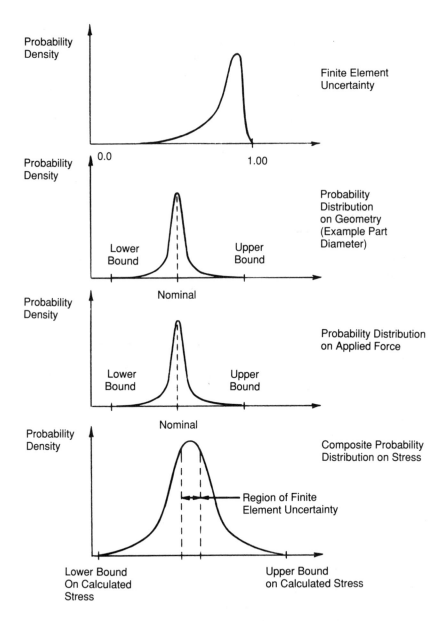

Figure 13.7 Probabilistic distribution of input parameters.

densities and comparing the results as a function of the density. It is known that all models should converge to the best result with increasing mesh density. This is, admittedly, a time-consuming procedure, but for cases in which many analyses are to be performed on similar components with one basic "production" model, it may be worthwhile to perform a convergence study. Even two runs will help to determine the model's sensitivity. If the mesh density can be doubled with a change of, for example, 1% in the results, then this is an indication that model is satisfactory and that both models are near the point of convergence.

Studies have been performed on analyses of steam turbine blades (2, 6) to determine the most cost-effective models in terms of number of elements and numbers of master degrees of freedom. This study is discussed in Chapter 11, Section 11.4.3. Figure 13.8 shows a basic model. The airfoil section of the blade is a geometrically complex shape with taper, twist, and compound curvature. The blade is modeled with three-dimensional solid elements with straight sides so that the model's accuracy will be affected by the number of nodes on the surface. The cross section of the airfoil was set at eight elements along the chord and two elements across the thickness. The number of elements along the length of the blade was varied and convergence of the natural frequency of the first three modes is shown in Figure 13.9. As expected, the lowest mode converged first, followed by the higher modes in order. The number of master degrees of freedom was also varied and the results followed the same general trend (Figure 13.10). Based on these results, it was decided that 18 layers of elements along the length and 36 master degrees of freedom gave the most cost-effective model. More important, this study also provided information on the error involved in future models using the standard gridwork.

13.2.4 Manufacturing and Operational Variations

Variations in actual dimensions must include the allowable tolerances and an estimation as to the probability of exceeding those tolerances. Operational data such as applied forces and wear of the parts must be obtained from the operating history of similar machinery and experience. Input data such as applied forces may be clarified by bracketing the data and looking at the sensitivity of the final result. Often, the exact force amplitudes, distributions, or frequencies may not be accurately known; however, an upper and lowerbound on each of the input parameters

Variable Number of Dynamic Degrees of Freedom

Variable Number of Element Rows Along Vane Length

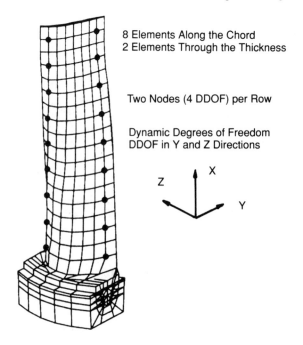

8 Elements Along the Chord
2 Elements Through the Thickness

Two Nodes (4 DDOF) per Row

Dynamic Degrees of Freedom
DDOF in Y and Z Directions

Figure 13.8 Steam turbine blade model used for parametric studies.

can usually be established. In these cases, two calculations may
be run for the upper- and lowerbound cases. The results should
then bound the problem, and if the worst-case results are still
acceptable, the analysis may be accepted on that basis.

The combined finite element uncertainty and manufacturing
variations can be shown, in principle, in Figure 13.11. The
curve represents the probabilistic distribution of an actual function, stress for example, which is to be predicted by finite element analysis. The finite element uncertainty is shown super-
imposed on the total uncertainty curve. If each of the input

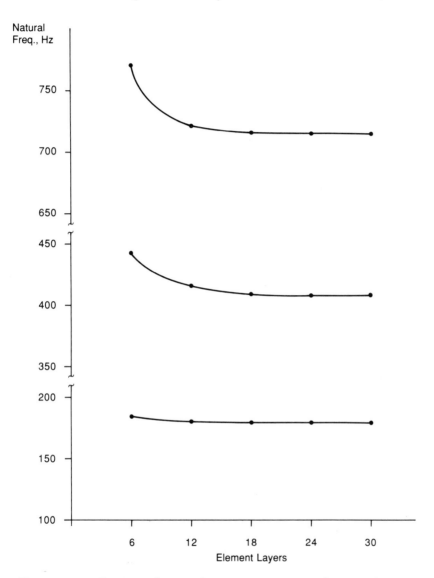

Figure 13.9 Number of vane elements vs. natural frequencies, 18 dynamic degrees of freedom.

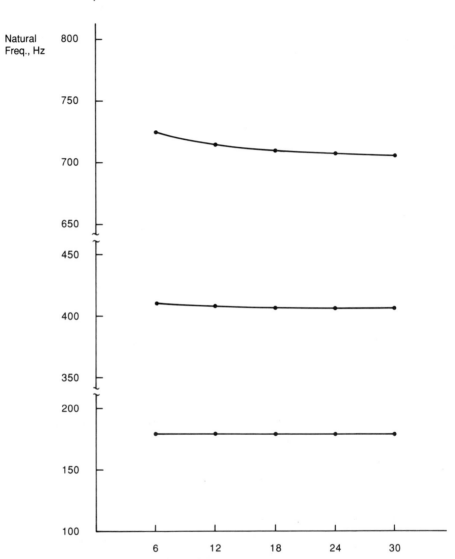

Figure 13.10 Number of dynamic degrees of freedom vs. natural frequencies, 30 layers of elements.

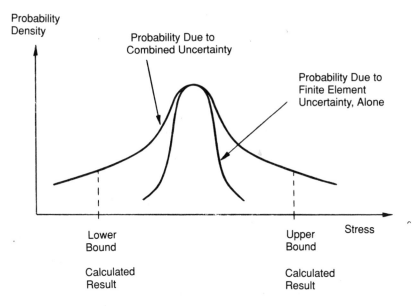

Figure 13.11 Composite distribution of finite element results.

parameters, including finite element modeling parameters, were bounded, the results can also be shown in Figure 13.11.

In some cases, it may be difficult to see the influence of input parameters at the beginning of an analysis. For example, a design analysis may have the objective of determining a component's susceptibility to fatigue that, in turn, is a combination of steady stress and dynamic stresses. The input parameters are steady force and dynamic force and its frequency. Not each of these three parameters can be quantified with the same degree of confidence, but not all of them will be as significant in determining the susceptibility to fatigue. By performing a series of finite element calculations using upper and lowerbounds on each of the parameters, it may be found that the frequency of dynamic loading, for example, is the most critical parameter because of a potential resonance problem. In that case, the magnitude of the dynamic stresses may be critical and further effort should be put into precisely defining the dynamic forcing frequency range.

13.3 SUMMARY

There is no automatic answer to the question of how accurate finite element analysis is. The accuracy will vary for each individual case. However, with an understanding of the basic behavior of the structure and some ingenuity reasonable, estimates can be given.

REFERENCES

1. MacNeal, R. H. and R. L. Harder, "A Proposed Set of Problems to Test Finite Element Accuracy," *Finite Elements in Analysis and Design*, No. I, North-Holland, Amsterdam, 1985.
2. Steele, J. M., N. F. Rieger, and T. C. T. Lam, "Development and Testing of a General Purpose Finite Element Model for Stress and Vibration Analysis of Steam Turbine Blade Groups," Joint Power Generation Conf., Oct. 1986, Portland, OR.
3. Haugen, E. B., *Probabilistic Approaches to Design*, Wiley, New York, 1968.
4. Wadsworth, G. P. and J. G. Bryan, *Probability Theory and Random Variables,*, McGraw-Hill, New York, 1960.
5. Steele, J. M., "Finite Element Modal Analysis of Steam Turbine Blading," 4th International Modal Analysis Conf., Los Angeles, Calif., Feb. 1986.

INDEX